W0225730

New Methods for the
Consolidation of Metal Powders

PERSPECTIVES IN POWDER METALLURGY
Fundamentals, Methods, and Applications

In Preparation
Volume 2
Vibratory Compacting—Principles and Methods

Volume 3
Iron Powder Metallurgy

Volume 4
Powder Metallurgy of Iron Alloys and Steels

PERSPECTIVES IN POWDER METALLURGY
Fundamentals, Methods, and Applications

Editors:

Henry H. Hausner	Kempton H. Roll	Peter K. Johnson
Adjunct Professor	Executive Director	Assistant Director
Polytechnic Institute of Brooklyn	Metal Powder	Metal Powder
Consulting Engineer	Industries Federation	Industries Federation

Volume 1

New Methods for the
Consolidation of Metal Powders

With an Introduction by Henry H. Hausner

Springer Science+Business Media, LLC 1967

The editors gratefully acknowledge permission to reprint the following:

Isostatic Pressing of Powdered Materials, by J. C. Jackson, *Progress in Powder Metallurgy* 20:159-167 (1964) (Copyright 1964 Metal Powder Industries Federation).

Hydrostatic Pressing of Powders, by C. E. Van Buren and H. H. Hirsch, *Powder Metallurgy*, pp. 403-440, Interscience (1960) (Copyright 1960 Interscience Publishers, New York).

Explosive Compacting of Metal Powders, by G. Geltman, *Progress in Powder Metallurgy* 18:7-13 (1962) (Copyright 1962 Metal Powder Industries Federation).

Consolidating Metal Powders Magnetically, by D. J. Sandstrom, *Metal Progress* (Sept. 1964), pp. 91-94 (Copyright 1964 American Society for Metals).

A New Method for Compacting Metal or Ceramic Powders into Continuous Sections, by F. Emley, *Progress in Powder Metallurgy* 15:5-13 (1959) (Copyright 1959 Metal Powder Industries Federation).

The Mechanism of the Compaction of Metal Powders by Rolling, by P. E. Evans, *Planseeberichte* 7:102-116 (1959) (Copyright 1959 Metallwerk Plansee).

The Compaction of Metal Powders by Rolling. I. The Properties of Strip Rolled from Copper Powders, by P. E. Evans and G. S. Smith, *Powder Metallury No. 3*, pp. 1-25 and 26-44 (1959) (Published by the Iron and Steel Institute and the Institute of Metals, London).

The Extrusion of Metal Powders, by N. R. Gardner, A. D. Donaldson, and F. M. Yans, *Progress in Powder Metallurgy* 19:135-141 (1963) (Copyright 1963 Metal Powder Industries Federation).

Hot Extruded Chromium Composite Powders, by R. V. Watkins, G. C. Reed, and W. L. Schalliol, *Progress in Powder Metallurgy* 20:149-158 (1964) (Copyright 1964 Metal Powder Industries Federation).

Beryllium Powder Forging, by N. G. Orell, *Beryllium*, pp. 102-119, *University of California Press* (1965) (Copyright 1965 Regents of University of California).

Slip Casting Metal Powders, by H. H. Hausner, *Proc. Metal Powder Industries Federation* 14:79-90 (1958) (Copyright 1958 Metal Powder Industries Federation).

Vibratory Compacting of Metal Powders, by J. L. Brackpool and L. A. Phelps, *Powder Metallurgy No. 7*, pp. 213-227 (1964) (Published by the Iron and Steel Institute and the Institute of Metals, London).

Library of Congress Catalog Card Number 66-22786

© *1967* Springer Science+Business Media New York
Originally published by Plenum Press in 1967.

Softcover reprint of the hardcover 1st edition 1967

All rights reserved

No part of this publication may be reproduced in any form without written permission from the publisher

Foreword

The rapid development of new methods for the compaction of powders during the last few years has resulted in new avenues in powder technology – especially in powder metallurgy – and new types of products have been developed with improved physical properties. These new methods for powder compaction have been described in a great number of publications, most of them printed as conference papers, or in a variety of technical journals. Should someone want to study these new methods of powder compaction, just to compare them, or to consider them for production purposes, it would be rather difficult to find the respective references. For this reason, Plenum Press and these editors have decided to collect some of the most instructive articles describing the new methods of compaction and to issue them as chapters of this book.

The first chapter of the book represents an introduction into the subject matter; it offers a discussion on compacting, a general evaluation of the new methods of compacting, discusses friction conditions during the various compacting processes, the powder characteristics and how they affect compacting. Attention is called to the effects of the method of compacting on the structure of the sintered material. The articles forming the content of the other chapters have been published previously in "Progress in Powder Metallurgy", Proceedings of Metal Powder Industries Federation Conferences (Chapters 2-I, 4, 6, 9-I, 10-II, and 13); in the British journal "Powder Metallurgy" (Chapters 8-II and 14); in "Planseeberichte fuer Pulvermetallurgie" (Chapter 7-I); in "Metal Progress" (Chapter 5); in the book "Powder Metallurgy" (New York, 1961) (Chapter 3-II); and the book "Beryllium: Its Metallurgy and Properties" (1965) (Chapter 12). Chapter 11 dealing with the latest development in powder compaction, was written especially for this book. There have been no previous publications on this method.

This book contains many facts and data, as well as a series of most stimulating ideas of interest to powder metallurgists and ceramists, and to practically everyone involved in techniques dealing with powders. It is the first volume of a new series of powder metallurgy books, each of which will deal with detailed information on a special aspect of powder technology. The editors and the publisher hope to satisfy, with this new series, the increasing demand for information about the rapidly growing fields of powder metallurgy, ceramics, and powder technology in general.

Henry H. Hausner
Kempton H. Roll
Peter K. Johnson

Contents

Chapter 1. Introduction

The Compaction of Metal Powders

Henry H. Hausner

Adjunct Professor
Polytechnic Institute of Brooklyn
and Consulting Engineer

During the last twenty years powder metallurgy has progressed in many new directions; new types of products have been manufactured, and new methods of manufacturing have been developed. The fabrication of many of the new products has been possible mainly due to the new fabrication processes.

The term "new fabrication processes" concerns the methods for powder fabrication, resulting in new types of powders; it includes also modified sintering processes which result in sintered materials with improved physical properties. The greatest, and probably most important of the powder metallurgy fabrication processes during the last few years, however, concerns the methods of consolidation of powders, the compacting of masses of powder particles into desired shapes.

Definition of "Compact" and "Compacting"

Before entering into a brief discussion of the new methods of compacting in this first chapter, and into quite some detailed discussions in the following chapters of the book, it might be worth

while to discuss briefly the term "compacting" in powder metallurgy terminology. Although the various glossaries or definitions
of powder metallurgy terms contain a description of the term
"sintering," one does not find any definition of "compacting." However, in most of these glossaries the term "compact" is defined.
In the "Standard Definitions of Terms Used in Powder Metallurgy,"
published in 1961 by ASTM (ASTM 243-61), the term "compact" is
defined as "An object produced by the compression of metal powder, generally while confined in a die, with or without the inclu-
sion of nonmetallic constituents."

In the "Definitions of Terms Used in Powder Metallurgy," published in October 1958 in the "Metal Powder Report" [Vol. 13(2)
(1958)], the term "compact" is defined as "An object prepared by
compressing powder in a mold or die."

"The Powder Metallurgy Glossary with Definitions" (Draft
Version No. 1) (Jan. 25, 1962), published by The Jernkontoret Powder Metallurgy Laboratory at the Swedish Institute for Metal Research, Stockholm, defines "compact" (Swedish "presskropp") as
"A body produced from powder by compaction." The term "compaction" is not explained in this glossary; however, "compacting
force" is defined as "The force directly applied to the work during
compacting," and the term "compacting pressure" is defined as
"The compacting force per unit area of the work projected perpendicular to the pressing direction."

The German term for "compact" is "Pressling," which indicates already its preparation by the application of pressure.

In the article "Discussion of the Term 'Compacting' in Powder Metallurgy," published in "Planseeberichte fuer Pulvermetallurgie" [Vol. 12:172-180 (1964)], attention was called to the fact
that some more recent methods for forming metal powders into
compacts do not necessarily require the application of pressure
(such as vibratory compacting and slip casting), and it was therefore proposed to define the term "compact" as "A configuration of
a mass of metal powder of well-defined shape and desired dimensions, characterized by density and/or strength higher than those
of the powder from which it is formed." The characteristics of
the compact also permit to define the term "compacting" as: "A
process for forming metal powders into well-defined shapes of de-

sired dimensions, during which the density and/or strength of the powder mass increases."

New Methods of Compacting

Approximately twenty years ago, the term "compaction," or consolidation of metal powders, usually referred to the method of unidirectional pressing of a powder in a hardened steel die, either by single action or by double action pressing. The presses used for this purpose were either of the mechanical or hydraulic type. The press capacity was usually the most important limiting factor in determining the size of the part, and most of the powder metallurgy parts were of rather small size.

During the last few years, a variety of powder compacting methods have been developed and are actually applied in powder metallurgy production. Table I lists eleven basically different compacting methods, eight of which are based on the application of pressure, whereas three methods are listed under compacting without the application of pressure. Within each of the listed types of powder compaction, many deviations are possible, and even more frequently several methods are combined for the fabrication of a powder compact, so that the number of methods for the compaction of powders is actually much greater than indicated in Table I.

These various methods for the compaction of powders, however, do not compete with each other, and it is hardly possible to apply one or the other method of compaction for one and the same part. Each of these powder compaction methods results in the manufacture of different types of products: different with respect to size and shape, as well as with respect to physical properties. Whereas the conventional, unidirectional pressure compacting is usually applied for rather small parts, the new compacting methods make it possible to consolidate relatively large masses of powder. Isostatic pressing may result in tungsten ingots of 2 to 3 ft in length and up to 15 in. diameter, weighing up to several thousand pounds. Powder rolling permits fabrication of sheets of, for example, 0.03 in. in thickness, 2 to 3 ft in width, and 50 ft or more in length. Extrusion of powders permits the manufacture of bars 10 ft long or much longer with simple or complicated cross section. Explosive compaction also permits fabrication of large

Table I.   Various Methods for Forming Powders
into Compacts

A. Forming of powders by application of pressure

    1. Unidirectional pressing
        a. Single action pressing
        b. Double action pressing
            1. Floating die body
            2. Mechanically moved die body

    2. Isostatic pressing

    3. Powder rolling

    4. Stepwise pressing

    5. Powder extrusion
        a. Powder direct
        b. Powder canned

    6. Powder swaging

    7. Explosive compacting

    8. Powder forging

B. Forming of powders without application of pressure

    1. Loose powder sintering in a mold

    2. Slip casting

    3. Vibratory compacting

---

parts, characterized by a high green density. A new avenue for
the manufacture of hollow and porous materials was opened when
the ceramic slip casting process was developed for metal powders.
Vibratory compaction can result in powder masses or parts of
very high green densities. A powder consolidated by vibratory
compaction can be perfectly densified by a subsequent extrusion
or swaging process. There are actually numerous combinations
of compacting methods in practical use.

Friction Conditions During Powder Compaction

During compaction of powders by any method listed in Table I,
a movement of the powder particles takes place, usually resulting
in friction between the particles, and also between the particles

Table II. Variables Affecting Friction Conditions
During Compaction of Powders

A. Powder characteristics

    1. Type of powder material
    2. Particle size
    3. Particle-size distribution
    4. Particle shape
    5. Particle surface

B. Characteristics of compacting device

    1. Method of compaction
    2. Material of compacting device
    3. Surface finish
    4. Atmosphere of compaction
    5. Temperature of compaction

C. Characteristics of the lubricant

    1. Type of lubricant
    2. Amount of lubricant
    3. Method of application
        a. Admixed with the powder
        b. Applied to the surface of the compacting device

and the wall of the device in which the compacting process occurs [1]. One distinguishes between (a) friction between the powder particles, and (b) friction between the particles and the die wall, or the wall of the equivalent compacting device [2]. The variables which affect these friction conditions are listed in Table II. One can distinguish between three groups of variables: those characteristic for the powder, for the compacting device, and for the lubricant applied during compaction.

In connection with unidirectional powder compaction, friction between powder particles and the die wall is one of the most important factors which affects the uniformity of density throughout the pressed part. Experience has shown that in this case the friction between powder particles themselves, sliding on each other, is a rather minor factor compared to the friction between the powder and the die wall, and that the latter is practically the sole source for the pressure decrease caused by friction loss.

The friction conditions are entirely different with other compacting methods. In powder rolling, for example, the powder particles have considerably more freedom to move, and therefore greater friction occurs between the particles; however, between the powder and the rolls which form the powder into strips or sheets, maximum friction is required, and any application of a lubricant on the surface of the rolls would be detrimental, and would result in a low density compact.

In isostatic pressing, hardly any movement of the powder particles along the flexible mold wall takes place, resulting in minimum friction loss. Between the powder particles a limited movement occurs; however, as already stated above, the friction loss on account of the movement between particles offers no special problems. In large parts to be pressed isostatically, however, the friction loss between powder particles may result in a decrease of density from the outside to the inside of the parts.

During the extrusion of powders, the particles again have great freedom to move, although to a lesser extent than in powder rolling; the friction between the powder and the wall of the extrusion die, however, is very strong, and offers many difficulties; it requires a special die design and also a special type of lubrication.

Table I shows that there are methods for powder compaction without the application of pressure. In any mass of powder friction exists between the particles. In the case of slip casting, however, the particles are separated by the liquid vehicle of the slip and move rather freely in the fluid. In this case, friction is at a minimum because the liquid vehicle acts as a lubricant. During vibratory compaction of a mass of powder, the particles also move and arrange to a higher packing density, but during this movement the particles are separated by the application of vibrations; during their movement they do not slide on each other, but rather jump, and overcome friction in this way.

Lubrication plays an important role in unidirectional compacting, not only during the compacting process, but also during the ejection of the pressed part from the die. Most of the other methods of compacting listed in Table I do not offer any special ejection problems, and therefore minor problems of lubrication during ejection.

Table III

Characteristics of a Powder Particle

A. Material characteristics

  1. Structure
  2. Theoretical density
  3. Melting Point
  4. Plasticity
  5. Elasticity
  6. Purity (impurities)

B. Characteristics due to the process of fabrication

  1. Particle size (particle diameter)
  2. Particle shape
  3. Density (porosity)
  4. Surface conditions
  5. Microstructure (crystal grain structure)
  6. Type and amount of lattice defects
  7. Gas content within a particle
  8. Adsorbed gas layer
  9. Amount of surface oxide
  10. Reactivity

Characteristics of a Mass of Powder

  1. Particle characteristics (see Table above)
  2. Average particle size
  3. Particle-size distribution
  4. Average particle shape
  5. Specific surface (surface area per 1 g)
  6. Apparent density
  7. Tap density
  8. Flow of the powder
  9. Friction conditions between the particles
  10. Compressibility (compactability)

The correlation between densification of the powder mass and the respective friction conditions is entirely different for each method of compaction, and it is the opinion of this writer that these correlations are not yet completely established, especially not for the more recently developed new methods of compaction.

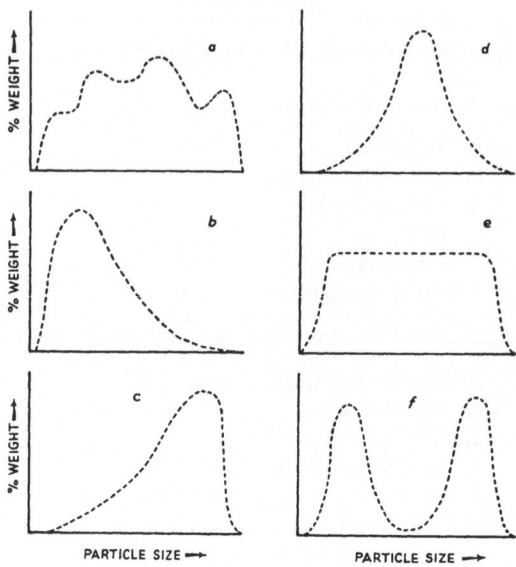

Fig. 1. Various types of particle-size distribution.

Powder Characteristics and Their Effects on the Compacting Methods

Powders exist with entirely different characteristics. The characteristics of a powder particle and of a mass of particles are shown in Table III. The question has frequently arisen as to whether there exist correlations between the powder characteristic and the compacting methods or, in other words, whether there is a special characteristic which is especially suited for one compacting method, but not for another. The answer is that such correlations definitely exist, but that they are at present investigated only in a very limited way.

There are compacting methods such as powder rolling, where finer particles are more advantageous than coarser ones. Other methods, for example extrusion or swaging, require elongated particles rather than equiaxed ones. For slip casting one prefers spherical particles rather than irregular or elongated ones.

It is to be assumed that the particle-size distribution plays a very important role in order to obtain satisfactory results in powder compaction. Figure 1 shows six different types of particle-size distribution. It is not yet known which particle-size distribution is best for each of the compacting methods listed in Table I. It is known, however, that for slip casting a particle-size distribution according to Fig. 1(b) offers advantages, whereas for vibratory compaction a particle-size distribution according to Fig. 1(f) with two maxima rather than with a maximum in one size is needed in order to obtain the desired high densities.

For several compacting methods where pressure is applied, a powder with high apparent density is usually favorable. Due to the fact that some pressure compacting methods require rather high friction between particles and the surface of the forming device, whereas others require the lowest possible friction, the surface configuration of the particles also plays an important role, in addition to their size and shape.

The flowing quality of a powder – whether it flows freely through a funnel, or only by application of vibrations, or not at all – is also one of the properties which are to be determined in automatic production using unidirectional pressure compaction. The flowing quality of the powder is also of importance in isostatic pressing and powder rolling, but to a much lesser degree in powder extrusion or forging, and of no concern in slip casting.

Attention is called to the existing but not well-known correlations between particle characteristics and the method selected for compaction in order to avoid possible disappointments in the selection of a powder for a compacting method for which it is not well suited.

Grain Structure of Sintered Materials as Affected by the Compacting Method

During compacting of metal powders into shapes, the powder particles are deformed and usually cold-worked during pressure compacting, but they are not changed at all during pressureless compaction. Deformation and cold working, however, determine the grain structure formed during sintering of the compacted material. The greater the deformation and cold working, the more crystal nuclei are formed and the smaller, therefore, is the grain size after sintering. The factors which determine microstructure

and grain size are very complex. It is evident, however, that
practically each of the compacting methods listed in Table I will
result in formation of a special grain structure of the sintered
material. Grain growth during sintering and the laws governing
the formation of the structure have been extensively described in
references [3] and [4].

The deformation of a powder particle depends on the plasticity of the powder material and the compacting method. More
uniform deformation of the particles in the powder mass occurs
in isostatic rather than in unidirectional compaction. Formation
of a structure of rather long fibrous grains during extrusion, and
of shorter fibered grains during swaging can be explained easily
by the mechanism of the two methods. Minimum orientation of
grain structure occurs by application of pressureless compacting
methods. Rolling of powders results in lesser orientation of
grains than the rolling of a solid material, because of the densification of the powder mass during rolling and the therefore lesser elongation of the grains within the particles. Tests have indicated that high deformation of particles in explosive compaction
results in the formation of a large number of crystal nuclei and,
therefore, should cause a smaller grain structure; however, the
small number of pores in explosively compacted powders contri-
butes again to faster grain growth and larger grains.

Actually, each method of compaction has to be evaluated together with the characteristics of the powder in order to predict
the type of grain structure which develops on subsequent sintering.

The following nine chapters are reprints of articles on new
methods of powder compaction published during the last few years.
The reader will receive some detailed information on new methods
of compaction which will be useful in the form presented, but probably will be very stimulating for further investigations in the complex field of powder compaction.

References

1.   H. H. Hausner and I. Sheinhartz, "Friction and Lubrication
     in Powder Metallurgy," Proc. 10th Annual Meeting, Metal
     Powder Association (1954), pp. 6-27.

2. P. M. Leopold and R. C. Nelson, "The Effect of Die Wall Lubrication and Admixed Lubrication on the Compaction of Sponge Iron Powder," Int. J. Powd. Met. 1 (3) : 37−44 (1965); 1(4):37-40 (1965).

3. H. H. Hausner, "Grain Growth During Sintering," in: Symposium on Powder Metallurgy, Special Report No. 58, Iron and Steel Institute, London (1954).

4. H. H. Hausner and R. King, "Effect of Powder Particle Size on the Grain Size of the Sintered Material," Planseeber. 8(1):28-36 (1960).

Isostatic Pressing of Powdered Materials

Harry C. Jackson

National Forge Company
Irvine, Pennsylvania

Introduction

Hydrostatic and/or isostatic pressing has been used for many years in forming materials.*

High pressures are playing an increasingly important role in development of new and higher properties in powder metals and ceramics.

During the last several years, there has been a great deal of research and development in this field. Both powder metallurgists and ceramists are beginning to take advantage of the uniform structure and controlled shrinkage in parts produced by this technique.

The unusual and extreme requirements of the nuclear age and modern industrial development have served as catalysis in the full realization of the potentials of this process.

---

*Originally promoted by H. D. Madden in 1913.

With the availability of faster and higher pressure pumping systems, large pressure vessels, and automation, the process is fast approaching the point of fulfilling requirements.

What Is Isostatic Pressing?

Isostatic pressing is the name used when applying pressure simultaneously and equally in all directions to a powdered material contained in a tightly sealed flexible mold.

For the purpose of this article, the following terms are defined. If the pressure media is a liquid then the process is more specifically referred to as hydrostatic pressing. However, if the pressure media is a gas then the term more commonly used is isostatic pressing.

A typical illustration of a hydrostatic or isostatic press is shown in Fig. 1.

The first step is filling the flexible mold with powdered material and sealing tight to separate the powder from the fluid. The mold is then placed in the pressure vessel. The pressure vessel closure is secured in place. The pumping system is started and the vessel pressurized to the desired pressure and then released. The vessel closure is opened and the pressed part removed.

The shape of the part pressed is determined by the shape of the mold. It is very easy to produce round, square, rectangular, or odd shapes. The reproduction of the mold is so sensitive that any imperfection or blemish on the inner surface of the mold will be reproduced on the compact.

Advantages of Isostatic Pressing

Companies are beginning to realize the potential of isostatically pressed parts. There is application for practically all types of powdered materials.

The advantages obtained from isostatic pressed compacts are listed below.

1. Uniform strength in all directions regardless of size.

2. Close dimensional and surface finish control.

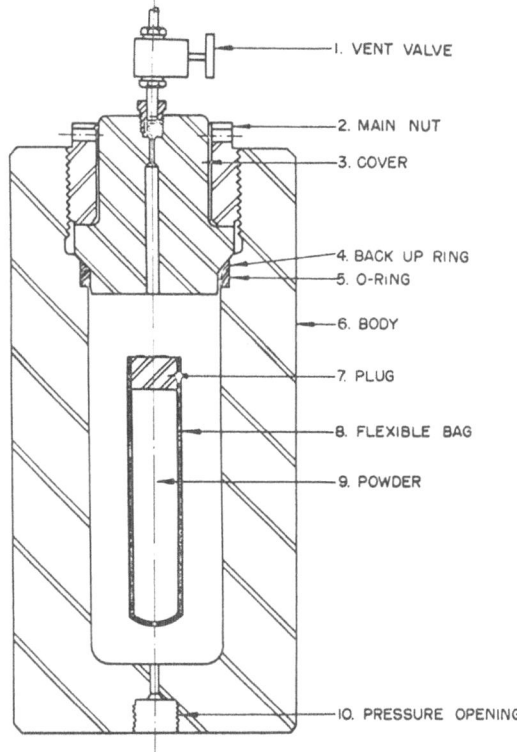

Fig. 1. Typical isostatic pressing vessel assembly.

3. Production of shapes that are impossible via other methods.

4. Completely homogeneous part without voids or air pockets and with reduced internal stresses.

5. Exotic and hazardous materials can be processed with minimum scrap and limited danger.

6. Greater ratios of length–to–diameter possible.

7. Excellent electrical properties.

8. Lower die cost through the use of rubber or plastic molds.

9. Lower cost of equipment and installation.

10. Space saving.

The Process

There are very few data concerning the results of this process. It is recommended that for specific applications, an evaluation program be set up to determine optimum operating parameters.

The pressure range varies for the powder used. Ceramic materials in most cases do not require pressure above 20,000 psi for optimum results.

Metal powders vary in pressures required from 30,000 psi to 100,000 psi with 30,000 psi to 50,000 psi the most common range.

Powder preparation plays a critical role in the final compact.

It is important that the materials flow properly and fill with uniform density without bridging within the mold.

Particle shape and size affects flow and compactability. Angular or irregular shaped powder particles offer the best change for making green compacts that are dense and strong.

Spray-dried materials offer very good results and fewer problems for ceramics.

Small amounts of moisture, up to 5%, produce higher densities. However, the higher the moisture content the more difficult it is to relieve air, resulting in laminations as well as cracking when baked.

Dimensional uniformity and straightness of the compact is directly dependent on uniform powder fill. If the powder will not give the proper fill, it may be necessary to vibrate, tamp, or pull a vacuum.

Powders having high compression ratio of $2\frac{1}{2}$ or more to 1 can cause difficulty. Vibration can be used to increase density of fill and reduce compression ratio. Also, the powder can be preformed and then placed in mold and pressed.

After the powder has been properly filled in the mold, it is important to seal the mold air tight. If leakage occurs in the mold during the compressing operation, the material may be spoiled. In some cases, a leak permitting the fluid to come in contact with reactive material has been known to form gases producing uncontrollable pressures which have damaged the vessel.

For research evaluation and large pieces, only one may be placed in the vessel at a time. However, when production is required, parts can be clustered and placed on a rack and placed inside the vessel. The pressure vessel can be prefilled with liquid to a predetermined level.

Precaution should be taken to install a vent valve at the top of the pressure vessel for the purpose of venting internal air. This is important for two reasons: one, it will decrease the pumping time necessary to reach a desired pressure and, two, because of its compressibility, entrapped air could cause the pressure vessel to become a missile if it were to rupture. After air has been vented, the valve can be closed and the vessel will be ready to pressurize.

Compacting pressure, dwell period, pump-up-time, and speed of venting will vary depending on type and size of material being pressed. This must be taken into account when considering production rates.

1. If the pressure is reached too fast and vented immediately, the material will not compact properly and have a soft core.
2. The dwell time plays an important part in that some materials require time to complete their compacting.
3. If the pressure is vented too fast, air entrapment may cause parts to crack or split.
4. Withdrawing air from bag will:
   a. maintain straightness and uniform compressing;
   b. eliminate sagging when held horizontally;
   c. cause bag to cling tightly to compressed part when pressure is released from vessel;
   d. reduce compression ratio.

The hydraulic pumping system will make up any pressure drop automatically that may develop because of the compacting of the material during the dwell time. Upon completion of the cycle, the closure may then be removed and compacted parts can be unloaded. The parts may then be removed from the bag or mold.

In many cases, dimensions can be so controlled such that no machining is required. Some materials can be machined while still in the green state and then placed in a furnace for sintering, while others are placed in furnaces for sintering first and then machined.

Type of Equipment

There are two types of isostatic pressing equipment available. They are known as "wet bag tooling" and "dry bag tooling."

Wet bag tooling is the most common and versatile. With this process the mold is loaded with powder, sealed, and loaded into the pressure vessel and pressurized. This process lends itself very well to laboratory, research and development, pilot plant, and high volume production work when properly tooled.

The advantages of wet bag tooling are lower cost of equipment, lower tooling cost, longer tooling life, and variety of parts that can be made at the same time.

Dry bag tooling utilizes the conventional presses used in dry pressing. The dies (molds) are fixed in place and made from rubber rigid enough to maintain its shape. The mold is filled with powder from hoppers automatically. The mold is closed by top punch of press. The isostatic pressure is then applied around outside of mold. Then the pressure is released, the mold returns to its original size. The part is removed by either the top or bottom punch.

The pressure vessel used in the wet bag process can be tooled in the same manner for dry bag process. See Fig. 5.

The advantages of dry bag tooling are the high production for small items, speedy powder loading, and lowered tool handling.

The disadvantages of dry bag tooling are inability to make complex shapes or more than one size or shape at a time, short tool life, higher cost for equipment, and limited maximum working pressure.

Isostatic Pressing Applications

The potential for isostatic pressing is much greater than industry realizes. Practically any type of powdered material can be compacted when properly prepared. Powder metallurgists are

finding nozzle inserts (dense tungsten parts) are being isostatically pressed to 92% theoretical density and hot forged to full density.
The high quality is the same regardless of size.

Materials being isostatically pressed are listed below.

Aluminum	Nickel
Aluminum oxide	Stainless steel
Beryllium	Solid fuel
Beryllium oxide	Tantalum
Carbides	Titanium
Carbon	Titanium – carbide
Cathode supports	Titanium oxide
Ceramics	Tungsten
Cermets	Uranium carbide
Columbium	Uranium dioxide
Graphite	Uranium – dioxide –
Iron	columbium – cermets
Magnesium	Uranium nitride
Magnesium oxide	Zirconium
Molybdenum	Zirconium oxide

Some of the items being isostatically pressed are listed below.

Balls for check valves, bearings, gauges, hole sizing, insu-
    lation
Blades or buckets for turbines
Carbide tools, rolls, and boring bars
Cermet tools
Compressed charges of explosive materials
Crucibles
Electrodes for consumable electrode vacuum arc melting
Exhaust nozzles
Fuel elements
Furnace boats
Grinding balls
Insulators
Nose cones for missiles
Radomes
Refractory spheres

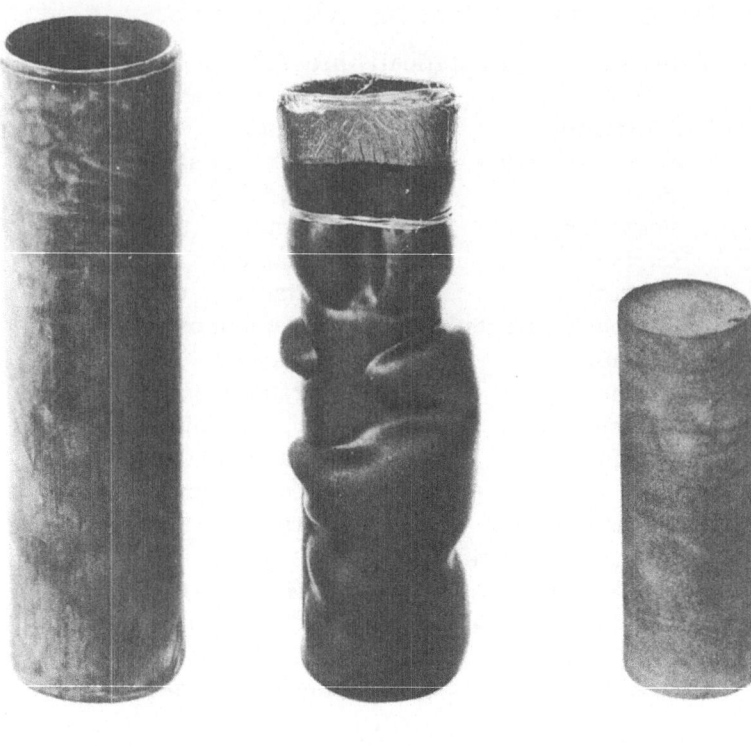

Fig. 2

Slab form billets for sheet rolling
Steel mill rolls
Susceptors (heating element not directly connected electrical-
    ly)
Thermocouple wire
Throat inserts for missiles

The question is often asked, what happens to the air in powder when it is compact. The photograph in Fig. 2 shows the
bubble when air has been trapped. The entrapped air does not
harm the mold shape.

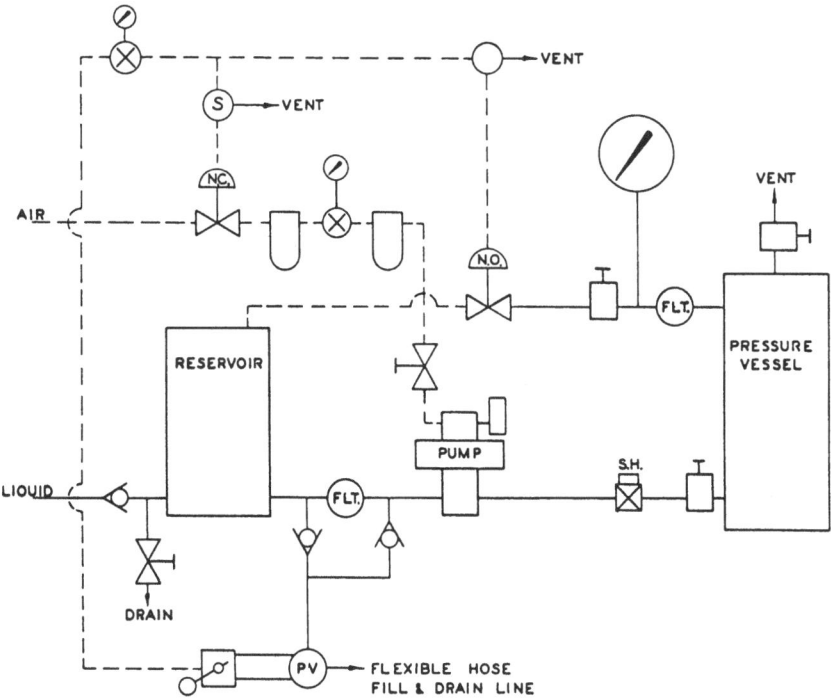

Fig. 3. Typical semi-automatic isostatic pressing system.

Hydrostatic Fluids

Liquids used in hydrostatic pressing can be any lubricating type not dangerous to personnel or equipment. When selecting a fluid, you must consider the following factors: compressibility, corrosion, lubricating valves, and compatibility with the mold tooling material. The three most common fluids used are water with rust inhibitor, glycerin, and hydraulic oil.

Isostatic Gases

Some systems are being used that require gas as the pressure medium. This involves compressors operating as high as 60,000 psi. In most cases where gas is used temperature is also involved. Temperatures to 650°F are externally applied, but for temperatures from 650°F to 3000°F, heating elements are generally placed inside the pressure vessel.

Fig. 4. Typical air-operated system.

Pumping Systems

In selecting a pumping system, consideration should be given to the type of application such as research, low production, high production, maintenance, and pressure range.

Standard pumping systems consist of the following: one or more high pressure pumps (air or electric), reservoir, filter, hand or automatically operated decompression valve, pressure gauge, rupture disc assembly, scavenger pump, hoist when required, console, and necessary valves and fittings.

Air-operated pumps have been used to advantage, as they will deliver a large flow, take up less space, cost less, and are easier to maintain. A typical air-operated pumping system and flow diagram is shown in Figs. 3 and 4.

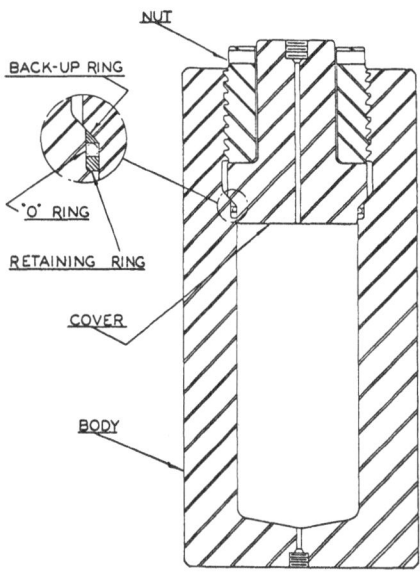

Fig. 5. Pressure vessel.

When selecting a pumping system remember that approximately 75% of the material compaction takes place in the first 3000 psi. Pumping systems can be furnished for manual operation permitting operator control at all times. In addition to the manual system, semiautomatic, completely automatic, or programmed automatic systems can be furnished as well.

Pressure Vessels

The pressure vessel should be constructed of high grade quality controlled steel. It is strongly recommended that the material used be ultrasonic and magnaflux tested.

It is standard practice to hydrostatic test the vessel 50% above rated operating pressure. Strain gage readings taken at the time the hydrostatic test is performed are very helpful in checking actual stresses against calculated stresses.

Good closure design is very important. The critical points to look for are tight seal, long life under continuous use, ease of opening and closing. There are two types of closures. The most

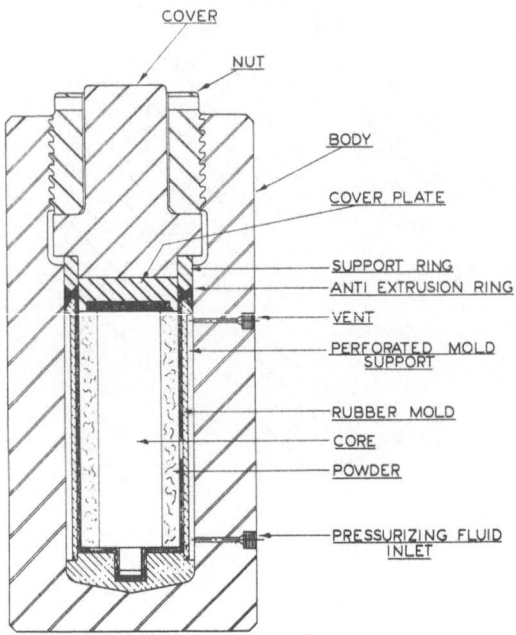

Fig. 6. Pressure vessel with dry bag tooling.

common is the continuous thread. This type is used for research and low production. It is low in cost but is slow to operate. For high production the breech type or interrupted thread is used. One-eighth turn of the nut allows the closure to be opened and closed. Typical vessel designs are shown in Figs. 5 and 6.

The safety of equipment can be designed to company, insurance, and state requirements. In most cases equipment can be designed to meet ASME Code. In all cases ample safety factors should be used.

Hoists

Vessel closures are sometimes very heavy and require mechanical means of removal.

Hoists of various types may be used. The main point to remember is to have a smooth, fine control. This helps to assure trouble-free life to threaded closure.

Air or hydraulic cylinders can be designed to open and close the breech closure and allow the hoist to remove without interference.

Production can be increased with the addition of cylinders to open and close the breech closure, a hoist to lift and return the closure, and a cylinder to swing the hoist and closure in and out of position by one push button control.

Equipment Selection

In selecting equipment, the following points should be kept in mind: (1) pressure rating of vessel and material, (2) size of vessel and, (3) operating speed of system.

Pressure Rating

When not familiar with pressure ranges required, it is advisable to have evaluation tests run. Each material may require a different pressure to compact. The particle size and shape, as well as the binder, plays a very important part in determining the pressure required. Lower pressure results in lower cost, less maintenance, and greater speed of operation.

Higher pressures result in higher green density and green strength. This allows the parts to be handled and machined in the green state with low scrap loss. Higher pressures also allow some materials to be pressed without a binder. Higher pressures do not always result in higher fired densities.

Size

Selection of the proper size pressure vessel depends on the maximum size part to be made, keeping in mind future as well as current requirements. Production requirements play an important part also. More than one piece at a time can be produced in larger vessels. It must be remembered when sizing a pressure vessel that the higher the pressure and larger the internal diameter, the higher the cost and greater time required to pressurize.

Keep in mind the maximum size bag tooling required. The bag tooling includes the mold, the support, and the clamps or other methods for sealing.

Operating Speed

The following factors influence operating speed and should be considered when selecting equipment: pumping system capacity, closure design, closure removal hoist, and bag tooling.

There are two types of pumping systems to consider. Air-operated pumping systems are less expensive and do a very good job to a certain point. However, the air-operated systems are limited on pumping volume and if high production is required, maintenance cost could be too high. It is advisable to purchase the more expensive electric motor driven pumps when higher production rates are required. You will have less maintenance with higher volumes, permitting higher production rates at lower cost.

There are three types of closures used. The first closure is the continuous thread type requiring threading and unthreading to operate. This type of closure is slow and not recommended for production. The second closure is the breech or interrupted thread type. One-eighth turn is all that is required to remove or close the closure. This type of closure is highly recommended for production. Another closure used is the plug or threadless type. This type utilizes a yoke and ram for the closure. This type is fast and allows for safer vessel design, because threads are eliminated and longitudinal stresses removed. The disadvantage with this type is the cost of press, space, and tooling required to load and unload the work.

Future of Isostatic Pressing

Isostatic pressing is being used on a limited basis throughout the world. Powder metallurgists and ceramists are beginning to see the advantages isostatically pressed items offer. Materials and parts that could not be used or made before have found new life. Every day new applications are appearing. Companies that have investigated isostatic pressing through laboratory units are now considering high volume rate production equipment.

As the demands increase so will the "know-how" in process and production equipment.

Chapter 3. Isostatic Compaction II

Hydrostatic Pressing of Powders

Charles E. van Buren and Harold H. Hirsch

General Electric Research Laboratory
Metallurgy and Ceramics Research Department
Schenectady, New York

Particulate matter can be consolidated into shaped bodies by the application of pressure in a fluid by a flexible, nonporous membrane. This process, referred to as "hydrostatic" or "isostatic" pressing, can be modified in a number of important ways which affect the final product produced. For example, pressure can be applied slowly or rapidly, the charge can be de-aired or not, and the powder may be made to load into the flexible membrane densely or loosely. These factors, along with the effect of pressure, particle hardness, size distribution, and granulations, were investigated to establish their effects on the quality of the pressed bodies and subsequent processing. This report presents the results of this investigation from which it has been determined that hydrostatic pressing produced compacts with improved mechanical properties versus die pressing, and that these properties are dependent on the above mentioned factors.

Introduction

Hydrostatic pressing, more conveniently referred to as "hydropressing," has been used for many years as a method for con-

solidating powders. The technique, until recent years, has been
largely applied to the compaction of ceramic powders, usually employed, except for a few important exceptions, for producing large
shapes. The significant literature and patents pertaining to this
process prior to 1958 are well documented in a paper by Newkirk
and Anicetti [1] and will not be repeated here.

During the last several years, there has been an ever increasing use of this technique for consolidating metal powders, as well
as an increased activity in the ceramic field. Powder metallurgists have "rediscovered" the method and are applying it to produce
large and complicated shapes from refractory metals and compounds and composite materials. Both powder metallurgists and
ceramists are beginning to take advantage of the uniform structure and isometric shrinkage that is produced in bodies prepared
by this technique. The application of the process is being further
stimulated by the availability of higher pressure systems and very
large pressing chambers.

This revitalized interest in hydropressing prompted the
authors to study the processing variables in a little greater detail
than undertaken by any single investigator hitherto. The results
obtained to date are summarized in this paper.

Description of the Process

The term "isostatic pressing" is the name given to any process that applies pressure simultaneously and equally in all directions to a powder mass. If the pressure transmitting medium is
a liquid, then the process is more specifically referred to as "hydrostatic pressing." If other pressure transmitting media, such
as rubber, plastics, powders, gases, are used the broader term
of "isostatic pressing" is employed.

The technique and equipment used in the current work is
shown in Fig. 1 and is quite representative of the industry. The
powder is loaded in a thin, flexible membrane, No. 9, usually
supported by some perforated or porous metal tube, No. 11. Since
the dimensional uniformity and straightness of the pressed body is
directly dependent on uniformity of powder fill, the powder is
tapped, vibrated, or tamped into the membrane. The shape of the
compress is determined by the shape of the membrane. Thus, it

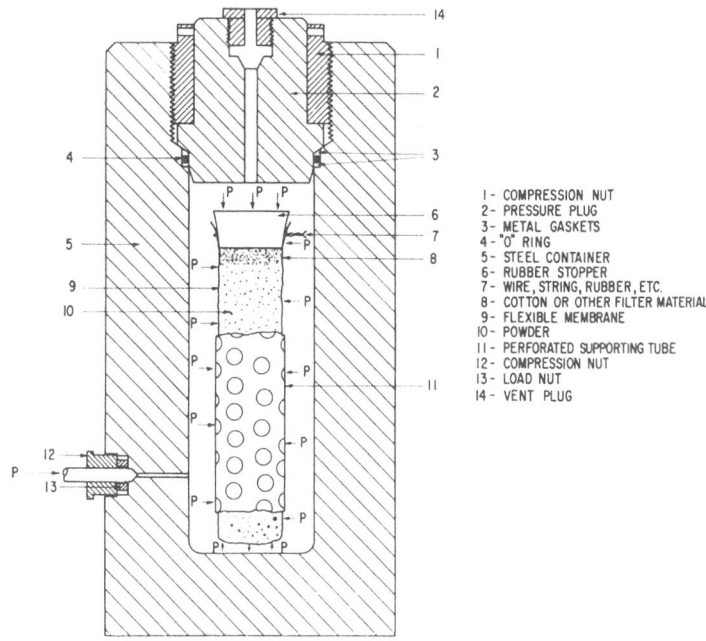

1 - COMPRESSION NUT
2 - PRESSURE PLUG
3 - METAL GASKETS
4 - "O" RING
5 - STEEL CONTAINER
6 - RUBBER STOPPER
7 - WIRE, STRING, RUBBER, ETC.
8 - COTTON OR OTHER FILTER MATERIAL
9 - FLEXIBLE MEMBRANE
10 - POWDER
11 - PERFORATED SUPPORTING TUBE
12 - COMPRESSION NUT
13 - LOAD NUT
14 - VENT PLUG

Fig. 1. Sketch of a typical hydro-pressing assembly. The pressure is supplied by an auxiliary pumping unit.

is very easy to produce round, square, rectangular, or fluted shapes. By way of illustration, the rather complicated iron rooster in Fig. 2 was born in the toy balloon shown alongside the rooster. Indeed, the reproduction of the membrane is so sensitive that any imperfections or blemishes on the inner surface of the membrane will be faithfully reproduced on the compress.

The closure of the bag is generally achieved with a rubber stopper by clamping or even gluing the bag to the stopper. It is frequently desirable to remove the air from the enclosed powder, and this can be readily accomplished by inserting a hypodermic needle through the rubber stopper and withdrawing the air with a vacuum pump. The small hole produced in the stopper is self-healing on withdrawal of the needle. To prevent clogging of the needle, cotton, No. 8, or some other filter material is placed on top of the powder. This cotton is entrapped in the end of the compress and is removed after pressing by cutting off this end. De-

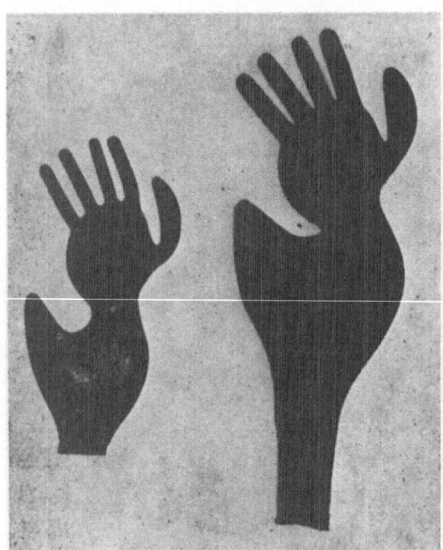

Fig. 2. An iron rooster prepared by loading iron powder in the toy balloon shown and hydro-pressing at 10,000 psi.

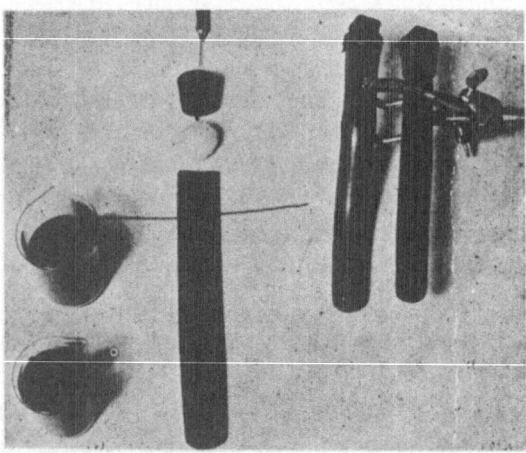

Fig. 3. An example of some of the items used in this program. From left to right: two types of iron powder, the rubber membrane, rubber stopper, hypodermic needle, cotton, and wire tie. De-aired and not-de-aired loaded bags are shown to the far right.

airing will help maintain straightness and uniformity of the compress for non-de-aired bags will be limp, whereas de-aired bags will be rigid and will not sag even if held horizontally by one end. The two specimens shown clamped at one end in Fig. 3 illustrate this point. The specimen to the extreme right has been de-aired and extends horizontally out from the clamp without any sag. The specimen second from the right, not de-aired, drooped until it touched the table.

The loaded bag is placed in the pressing chamber No. 5 (Fig.1) (this can be horizontal or vertical), filled with the pressing fluid, and then the chamber is sealed with a plug and jacket arrangement, one type being shown in Fig. 1. The vent plug No. 14, permits the removal of the small amount of remaining air, a desirable procedure, since compressed air provides dangerous potential energy in the event of container failure. Pressure is supplied to the chamber through the side orifice leading from some sort of high pressure pump.

The loaded bag is now subjected to equal pressure from all directions and powder compression occurs axially as well as longitudinally. No wall friction exists such as in die pressing, only internal particle friction, which must be very low for the compress, has a very uniform density throughout. Indeed, it has been claimed that with a given powder and compacting pressure, identical densities will be obtained with compresses varying from 1 to 10 in. in diameter [2]. Without a supporting tube, the loaded bag could virtually touch the container wall, so that full use of the container can be made.

Upon release of the pressure, the membrane will often spring away from the compress if no de-airing was performed, whereas in de-aired specimens, the bag will cling tightly to the compress until the vacuum is broken. Thin-walled membranes, especially with powders of high compression ratios, will occasionally wrinkle and introduce surface imperfections in the compress. Sometimes, the membrane will be pinched into the compress and must be torn to remove it, or it may be stuck to the compress especially if coarse powders are used. In most cases, these problems can be overcome by granulation, by using heavier or tougher membranes, and by coatings on the inner surfaces of the membranes so that the reusable life of the membranes is frequently 10 to 20 or more pressings if the pressures are low.

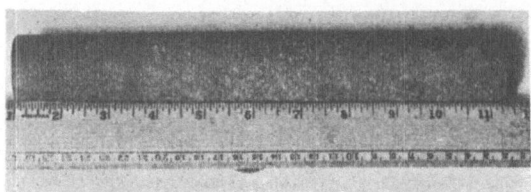

Fig. 4. An as-pressed molybdenum rod after removal of
membrane.

The finish of the hydro-pressed part can be quite good, as illustrated by the surface of the as-pressed rooster in Fig. 2. Fine
powder, smooth membranes, and good loading techniques are important. With good care, reasonable straightness can be achieved,
as shown in Fig. 4, a hydro-pressed molybdenum rod. Generally,
though, for good finishes and tolerance, some machining must be
done. This frequently can be carried out on the green compress,
for green strength is surprisingly high even without a binder and
shrinkage is very uniform and reproducible.

The Experimental Program

The Purposes

The major purposes of this study were to establish the rela-
tionships between the green and sintered properties of powders
and their forming pressures. The influence of de-airing of the
powder on these properties was also evaluated. Some of the factors distinguishing die-pressed products from hydro-pressed material were also investigated. It was the intention of this program
to produce some generalizations concerning the more favorable
conditions for hydro-pressing different classes of powder.

The Procedure

For the purpose of this program, three general types of powders were chosen: (1) a ductile metal – iron, (2) a nonductile metal – tungsten, and (3) a single component ceramic – aluminum
oxide. Further, to determine the effect of particle size, each
material was studied using a fine and a coarse powder and an intermediate size in the case of iron and tungsten. The characteristics of the powders are itemized in Table I. It is clear from the

Table I.   Characteristics of Powders Investigated

Classification	Size range Tyler Sieve No.	Apparent density theoret., %	Flow rate, sec/50 g	Micro-hardness[a]
	Electrolytic Iron Powder[b]			
Fine–A-230	100% −325 mesh	32.5	None	—
	90% 10–44 μ			
Medium–A-281	22% −325 mesh	42.5	27.7	94
	78% −65 + 325 mesh			
Coarse–242	100% −42 +100 mesh	49.0	21.5	99
Granulated fines	70% −325 mesh	35.7	None	—
	30% −16 +325 mesh			
	Tungsten Powder[c]			
Fine	100% −10 μ	9.5	None	—
	90% 0.8–10 μ			
Medium	100% −35 μ	17.6	None	—
	90% 2.3–35 μ			
Coarse	100% −42 +100 mesh	41.8	14.8	250
Granulated fines	70% −16 +100 mesh	20.0	None	—
	24% −100 +325 mesh			
	6% −325 mesh			
	Aluminum Oxide Powder			
Fine[d]	0.2–1.0 μ	9.5	None	—
Coarse[e]	33% −100 +150 mesh	38	57.8	2400
	67% −150 +270 mesh			
Granulated fines	65% −16 +35 mesh	17.4	None	—
	35% −35 +270 mesh			

[a] The Vhn obtained on powder particles with 50 g load for iron, 300 g load on tungsten, and 100 g load on alumina.
[b] Crane Metal Corp. designation.
[c] General Electric Co.
[d] Linde Co. (A Grade).
[e] Norton Co. (38 Grade).

particle size analyses that the fines, mediums, and coarse are not the same for all three powders, but rather represent what is generally accepted for these classifications for each of the powders. The granulated fine powder was obtained by hydro-pressing the original fines at 5000 psi and then abrading the compress through an 18 mesh screen. The apparent density and flow rate are extremely important in the hydro-press process. The higher the density, the larger the compress that can be produced with a given container size, for the powder will then have a smaller compression ratio; that is, less volume change will occur upon press-

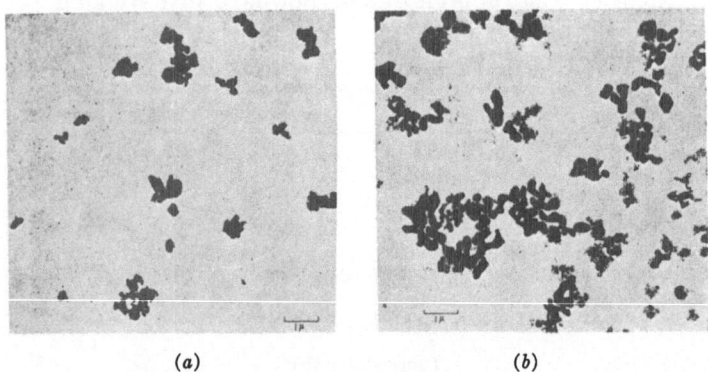

(a)   *(b)*

Fig. 5. Electron micrographs showing the powder contours. (a) Fine tungsten and (b) fine alumina. 5000 ×.

ing.   Proper flow rate aids rapid and uniform bag filling and con-
tributes to dimensional uniformity.   Granulation was found to be
beneficial to both of these factors.

Figures 5 and 6 show the general shape of the powders used
and provide additional help in analyzing their behavior.

Specimen Preparation

The metal powders were compacted into test specimens by
hydro-pressing and by die pressing.   The hydro-formed specimens were prepared by merely pouring the material into a rubber
mold which in most cases measured 1 in.^{2} by 9 in. long with a
0.015-in. wall thickness (Fig.4).   A piece of absorbent cotton was
then placed in the bag on top of the powder to prevent the powder
from being drawn from the bag during the de-airing (evacuation)
process.   The loaded bags were then sealed with a rubber stopper
and, where desired, evacuated by inserting a hypodermic needle
through the stopper.   The specimens were then hydro-pressed at
2,5,12.5,30,and 50 tsi in a hydrostatic press having a $1\frac{1}{2}$-in.bore,
9 in. deep.   Full pressure was reached in 5 min and held for 1
min.   By simply shutting off the pump, the pressure was released
and the specimen, still enclosed in its bag, was removed. The rubber mold did not adhere to the compresses at the lower pressures
but at 30 to 50 tsi the mold had to be peeled off the specimen.

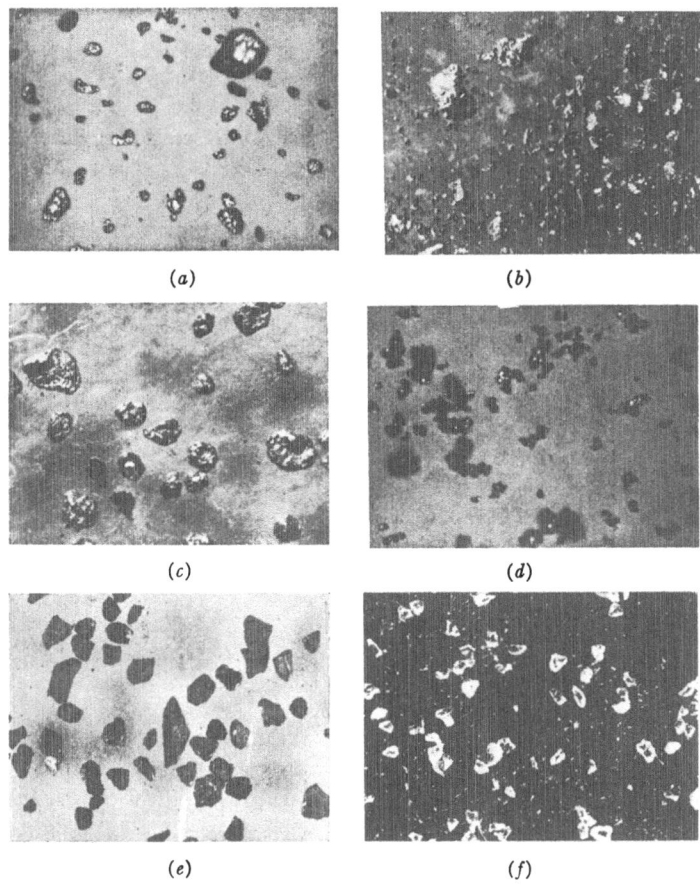

Fig. 6. Photomicrographs showing the powder contours of the coarser powders. (a) Fine iron (125 ×), (b) medium iron (25 ×), (c) coarse iron (13 ×), (d) medium tungsten (125 ×), (e) coarse tungsten (10 ×), and (f) coarse alumina (13 ×).

For pressures up to 5 tsi, a larger unit having a bore of 10-in. diameter by 18 in. deep was available (Fig. 7). * This unit employs the pressure seal depicted in Fig. 1. The 100,000 psi unit* employs the very simple seal shown in Fig. 8C. This sketch shows a few other common seals. In Fig. 8D, two variations of a

---

* Both hydro-pressing units were purchased from the High Pressure Equipment Co., Erie, Pa.

Fig. 7. A commercial hydro-pressing unit capable of
reaching pressures of 10,000 psi.

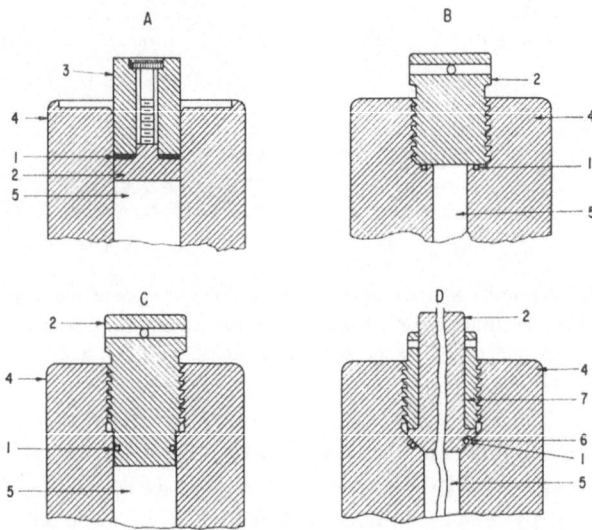

Fig. 8. An example of some of the typical pressure seals for
hydro-pressing chambers. The number parts are: 1) rubber
gasket or O ring; 2) sealing plug; 3) compression piston; 4) container; 5) pressing chamber; 6) leather gasket; 7) compression
nut.

type of seal developed by the Watervliet Army Arsenal, Watervliet, New York, are illustrated. This combination gasket of rubber and leather can withstand pressures of 200,000 psi. Note that for all these chambers, buttress threads are recommended for the screw parts. Sharp corners are to be avoided.

In Fig. 8A an arrangement is shown that permits one to use a standard pressing die and any hydraulic press to obtain hydrostatic pressing. The compression punch, No. 3, when loaded by the press platen, transmits the load to the hydraulic fluid in the chamber, No. 5. A Bridgeman seal is shown which insures positive sealing of the rubber gasket by providing greater pressure on the gasket than exists in the chamber through the unsupported section in the center of the sealing plug, No. 2. A fluid well is shown on top of the chamber. The authors have successfully used this type of equipment up to 100,000 psi.

In order to measure green density and strength, the as-pressed bars were machined into round test pieces, $\frac{1}{2}$-in. diameter for alumina and 0.4 in. for iron and tungsten. The alumina and tungsten specimens were made $1\frac{1}{4}$ in. long, whereas the iron samples were 2 in. long to provide sufficient length for the tensile tests on the sintered bars. It was not always possible to machine the green pressed bars below $12\frac{1}{2}$ tsi forming pressure, especially with the coarser powders. The test specimens were cut longitudinally from the compress. The transverse rupture measurements were made on 1-in. spans using 3 point loading and a loading rate of 0.01 ipm. Measurements in most cases were made in triplicate or quadruplicate, with at least two pressings used for each set of conditions.

The sintered density and strength in the case of tungsten and alumina were determined on test bars machined to shape prior to final sintering. The lower pressure specimens were prefired, that is, fired at low enough temperature to strengthen the compress without introducing noticeable shrinkage. This permitted machining without breakage. The iron specimens were machined after final firing into round test specimens having a gage section of 0.113-in. diameter and a $\frac{1}{2}$-in. gage length (Fig. 9) and strained at the rate of 0.01 ipm. Again, triplicate or quadruplicate specimens were tested from at least two pressings for each condition.

Table II. Conditions for Sintering Compacts

Powder	Sintering temp. °C	Time to temp.	Time at temp., hr	Atm	Setter material
Iron	1150	1 hr	1	Hydrogen	Placed on alumina grain
Tungsten	1850	To 700°C—8 hr 700–1850°C—6 hr	2	Hydrogen	Embedded in alumina grain
Alumina	1750	To 1300°C—10 hr 1300–1750°C—6 hr	2	Air	Placed on alumina grain

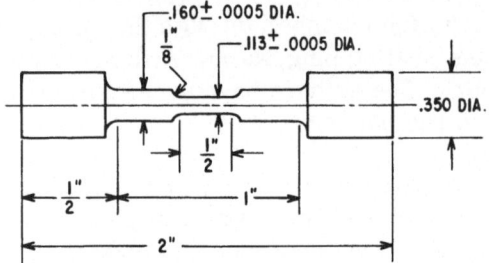

Fig. 9. Dimensions of specimens used for the tensile strength measurements of iron.

The sintering conditions are summarized in Table II and were used for both the hydro-pressed and die-pressed materials.

In order to obtain a comparison of the density and strength of powder consolidated in a die versus the same powder consolidated by hydro-pressing, the aforementioned powders were compacted in a hardened rectangular split die measuring $\frac{1}{2}$ in. wide by $1\frac{1}{2}$ in. long. Compaction was performed single acting at $12\frac{1}{2}$ and 30 tsi with a slow action 100−ton press. Sufficient powder was used to produce bars $\frac{1}{2}$ in. thick at each pressure and a wax lubricant was applied to the walls of the die. With the exception of iron, the pressed bars were machined into $\frac{3}{8}$-in. rounds by $1\frac{1}{2}$ in. long prior to sintering. This length was adequate for the transverse tests on alumina and tungsten, but a $2\frac{1}{4}$-in.-long specimen was required for the tensile tests with iron so a somewhat larger die was employed. The transverse bars were tested as-sintered, whereas the iron bars were machined in accordance with the dimensions shown in Fig. 9.

It was also thought to be of some interest to put some numbers on the density variation known to exist in die pressing. For
this purpose, a $^3/_4$-in. bore die was used to produce pieces $1^1/_2$ in.
long providing an L/ D ratio of 2 to 1. A wax wall lubricant was
employed and the compresses were made using the die with only
the top punch moving (single acting). One-quarter-inch thick
slices were then cut transversely out of the compresses and their
densities determined by weight and dimensions.

Discussion of Results

The behavior and trends of the hydro-pressed powders as a
function of forming pressure, particle size, and de-airing were
evaluated on green and sintered compacts. In the case of iron and
tungsten, the majority of results were obtained on the powders
with de-aired loadings and only check points made on not-de-aired.
The reverse procedure was used with alumina, i.e., the majority
of samples not-de-aired with check points with de-aired loadings.

Green Properties

Iron

Iron powder compacts can be hydro-pressed to produce green
density varying from 50 to 90% of theoretical density (7.87 g/ cu
cm) with forming pressures ranging from 4000 to 100,000 psi (Fig.
10). As the pressure is increased to 60,000 psi, the density of the
compacts rises rapidly from 50 to 80% of theoretical density. The
densification of the compacts above 60,000 psi is still significant,
and even at 100,000 psi the density curves are not leveling off. The
green strength of these compacts also increases very rapidly and
more uniformly than the density with increasing pressure. As with
the density, there is not any indication of a leveling off and it appears that the transverse rupture strength of 7800 psi obtained
with 100,000 psi forming pressure on the fine iron powder could
be exceeded with higher forming pressures.

It has also been found that as the particle size of the material
being hydro-pressed is increased, the resultant green density also increases. This improvement, however, is not reflected in a
corresponding increase in green strength. As can be noted, the
coarser the powder the weaker was the resultant compact. Indeed,

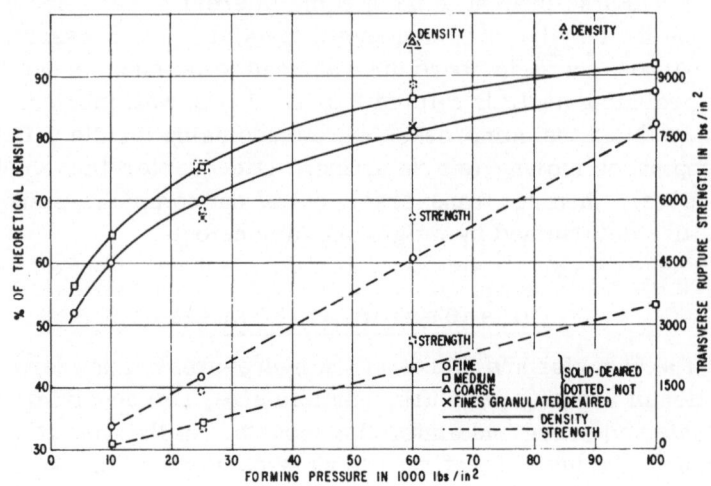

Fig. 10. Green density and green strength of hydro-pressed iron powder versus forming pressure.

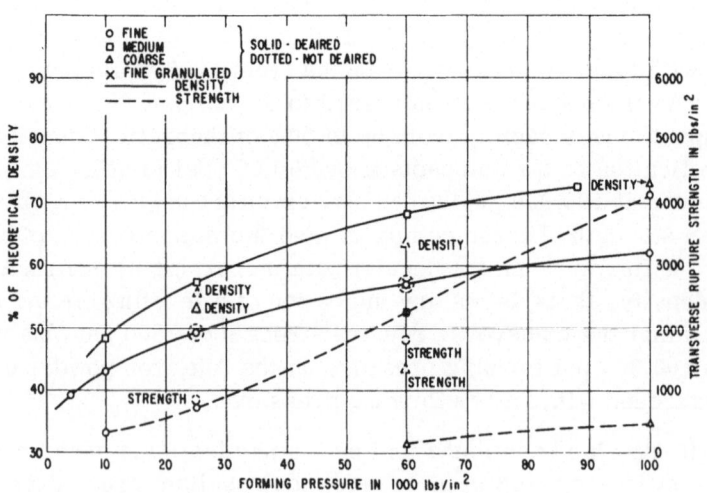

Fig. 11. Green density and green strength of hydro-pressed tungsten powder versus forming pressure.

it was impossible to obtain good green transverse test bars with the coarsest iron powder. Compacts made of fine iron powder are 120% stronger at 100,000 psi than those made of medium-sized iron powder even though the latter has the higher green density.

The green density and strength of iron powder compacts were affected very little by not de-airing at the check points of 25,000 and 60,000 psi (the dotted points). It is interesting to note, though, that when compared to the de-aired data, the not-de-aired values are lower at 25,000 psi and higher at 60,000 psi. This fact is of little value as far as the green density is concerned but might be of great value for the green strength. For example, if the same straight-line relationship exists for the not-de-aired strength versus pressure as does exist for the de-aired, one could reasonably expect that significantly higher green strengths could be obtained by not de-airing with high forming pressures. At 60,000 psi, the not-de-aired strength is 23% higher than the de-aired strength.

The coarse iron powder could not be hydro-pressed in one piece below 60,000 psi when de-aired, but when not-de-aired pressures as low as 25,000 psi could be used. Although these compacts could be handled by using gentle treatment, they were considerably weaker than the medium powder and therefore no attempt was made to measure their green strength.

Granulating the fines did not produce much change in the green density or strength, and therefore it is only beneficial when it is desirable to produce larger compacts with a given piece of equipment, for a greater mass of granulated powder can be put into a given bag than with ungranulated material.

Tungsten

The green density of tungsten compacts increased more uniformly over the pressure range of 4000 to 100,000 psi than in the case of iron, reaching a maximum value of 73% of theoretical density. This is considerably lower than the value of 93% obtained with iron and from the slope of the curve (Fig. 11) it appears that super-high pressures would be required to approach values of the order of 90% of theoretical density.

The green strength of tungsten also increases with pressure, but, as with the density, at a much lower rate than observed with

iron powder. The maximum transverse green strength obtained
is 4100 psi at 100,000 psi forming pressure, which is 46% less
than the maximum for iron powder.

The particle size of the tungsten powder did not have the same
effect on the green properties as was observed with iron. With
tungsten the medium and coarse powders produce denser com -
pacts than the fine, while the medium has the advantage over the
coarse. There is also a bigger spread between the green density
of the medium and fine tungsten than there is for medium and fine
iron, even though the latter shows a bigger spread in particle size.
At 88,000 psi this spread amounts to $11\frac{1}{2}$% as compared to $4\frac{1}{2}$%
for iron. As with the iron powder, the green strengths of the
tungsten compacts were greatly reduced as the particle size of
the powder was increased. The coarse tungsten was a little
harder to handle than the coarse iron, thus indicating poorer bonding of the particles under pressure. These results are not surprising, since tungsten, being much less ductile and stronger than iron,
will show considerably less flow and cold welding at particle-to-
particle contact points. Also, the photomicrograph (Fig. 6E) of
the powder shows the coarse powder to have fairly smooth sur.-
faces.

Not-de-airing had no effect on the green density of the fines
but it reduced the strength 20% at 60,000 psi forming pressure. At
25,000 psi, the strength of not-de-aired bars was higher than the
de-aired samples by a very small margin. The density of the
coarse tungsten was improved slightly by not de-airing, while the
strength was increased ten-fold. Granulation, on the other hand,
had no effect on the density or strength.

A direct comparison of the green properties of iron and tungsten shows that the medium tungsten has a green density 17% lower
than that for fine iron (these two grades are more nearly equivalent than the two fines) at the same forming pressure. The strength
of the tungsten under the same conditions is 48% lower than the iron,
thus indicating that ductile materials will yield higher green properties than hard, brittle materials.

Aluminum Oxide

The same results were not realized with hydro-pressed ceramic materials for, as would be expected, it has been found that

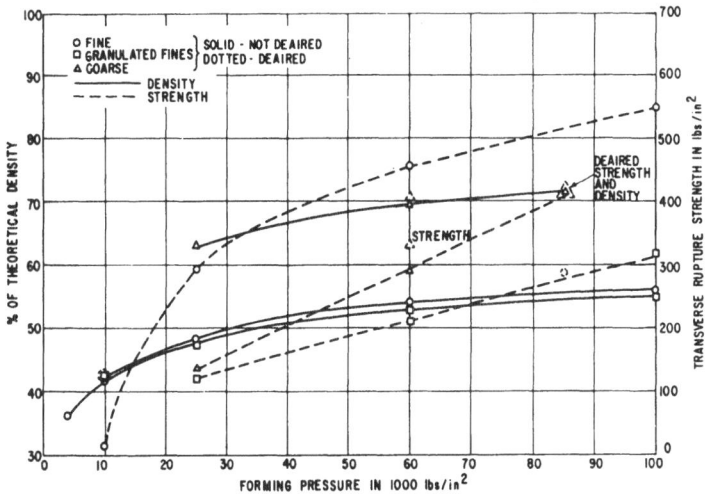

Fig. 12. Green density and green strength of hydro-pressed alumina
powder versus forming pressure.

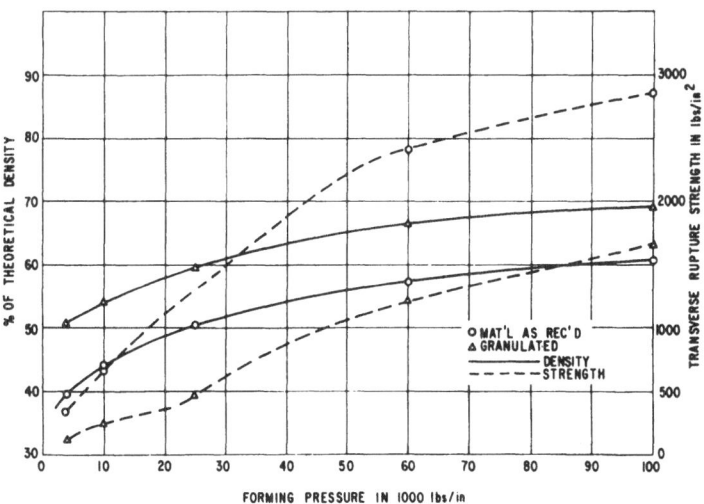

Fig. 13. Green density and green strength of hydro-pressed forsterite
powder versus forming pressure.

the green properties are generally lower than those obtained with metal powders.   The green density of both the fine and granulated fine alumina (Fig.12) is practically the same over the entire pressure range. Almost all the densification of this material takes place below 30,000 psi with only minor improvements above 30,000 psi forming pressure. The same is true with the coarse alumina, although its green density is 29% higher than the fines at 30,000 psi.   The spread between the fine and coarse densities is just about the same over the whole pressure range studied.

The green strength of the fine alumina increases rapidly between 10,000 and 25,000 psi and then rises almost linearly at a lower rate as the forming pressure is increased to 100,000 psi. The strength of the coarse alumina also rises linearly between 25,000 and 85,000 psi and is 51% lower than the strength of the fine alumina at 30,000 psi.   The gap between the strengths of the two powders decreases as the forming pressure is increased, suggesting that possibly cold welding of alumina can be made to take place, or it could be due to reduction of the coarser particles through crushing under load. Surprisingly, the strength of the granulated fines is 64% lower than the ungranulated material at 30,000 psi and the gap between the two remains essentially constant over the pressure range of 30,000 to 100,000 psi.

While de-airing had no effect on the green density of the fine or coarse powders, it did tend to increase the strength of the coarse at 60,000 psi but not at 85,000 psi.

The coarse alumina reacted the same as the coarse metals in that it could not be pressed in one solid piece below 25,000 psi and the compacts made at the higher pressures were extremely weak and had to be handled with care.

Since this material might be classified as a brittle ceramic, and was chosen as a representative oxide for comparison to the tungsten, it is of interest to see what relationship exists between their green properties. It can be noted that while the maximum per cent of theoretical density obtained with alumina is only 10% lower than for tungsten, the green strength is 85% lower. This indicates, as would be expected, that there is a much stronger bond between the metal particles than there is between the oxide particles assuming that the particle shapes are similar.

A cursory study was made with a multi-oxide ceramic classified as a forsterite $(2MgO \cdot SiO_2)$. The raw batch comprises two materials, both of which are considerably "softer" particles than alumina. The material used has a bulk density of 0.462 g/cu cm, which is 14.3% of theoretical density. No attempt was made to investigate the effect of particle size, except by the means of granulation. From the few results obtained with this material, it was found that there are apparently significant differences in the hydro-pressed properties of different ceramic materials, just as there are with metal powders.

As shown in Fig. 13, the green density of this material is almost as high at any given pressure as that obtained with coarse alumina, although its particle size is considerably smaller. It also follows the same pattern as the alumina in that most of the densification takes place below 30,000 psi forming pressure and higher pressures do not result in any significant increases. Furthermore, while granulation did not affect the green density of alumina, it produced higher densities with this material to the extent of 10 units over the entire pressure range. At 30,000 psi forming pressure, the granulated density is 17% higher than the material as-received. The green strength increases at a rapid rate up to 50,000 psi forming pressure and then at only one-fourth of this rate in the next 50,000 psi. The green strength at 100,000 psi is five times that of fine alumina.

Granulation, however, produced a 60% reduction in the green strength of the as-received material at 30,000 psi. Over the range of 60,000 to 100,000 psi, the gap between the strengths of the two materials is relatively constant, amounting to about 1200 psi. A similar effect was obtained with alumina.

Sintered Properties

Good green properties are of little value other than for handling and machining unless they are reflected in improved sintered properties. Therefore, the green bars were sintered at temperatures, times, and atmospheres that were selected as being representative for each material (Table II). It was realized that these do not represent the optimum conditions for heat treating these materials, but it was felt they should at least indicate the effects of forming conditions and particle size on sintered properties. It

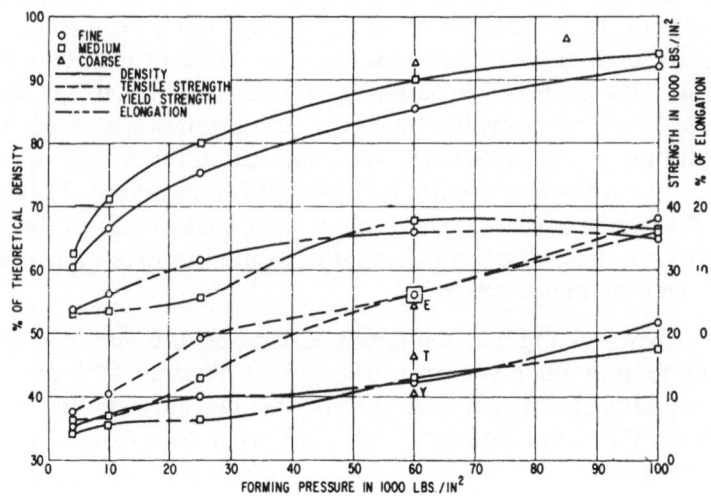

Fig. 14. Sintered density and sintered strength of hydro-pressed iron powder versus forming pressure.

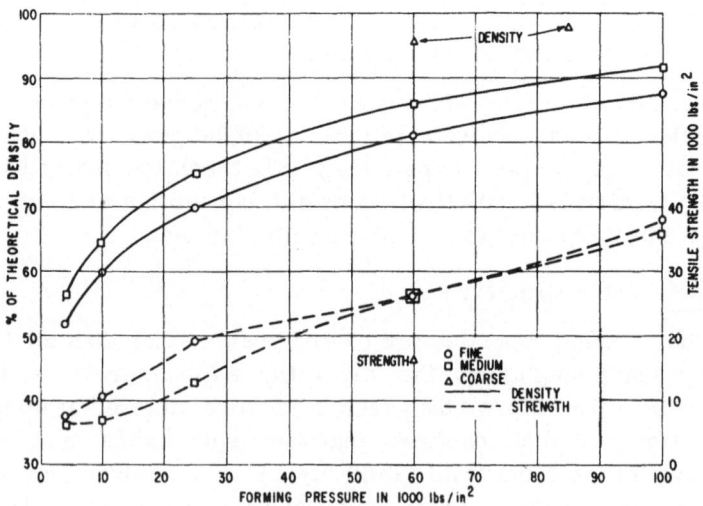

Fig. 15. Green density and sintered strength of hydro-pressed iron powder versus forming pressure.

was also realized that the rate of heating to temperature is extremely important, especially when the green density of the compacts becomes high, for fewer interconnected pores are available
for gas release, and earlier close-off of pores is possible. This
condition, which becomes more severe with hydro-pressed material because of the higher densities produced but possibly less
severe because of the ability to de-air the powders, suggests a
study of sintering conditions. Unfortunately, time did not permit
this kind of study but a few bars of alumina were heated at a faster
rate than indicated in the table, and will be discussed a little later.

As indicated in the table, the tungsten and alumina specimens
were heated to the sintering temperature at a slow rate, while the
iron specimens were brought to temperature at a fairly fast rate.

Iron

The sintered density of fine and medium de-aired iron (Fig.
14) increases as the forming pressure increases in much the same
manner as did the green density. The density of the coarse pressings showed essentially no change upon sintering, whereas the medium and the fine compacts increased 3-4% and 4-5%. This nar-
rowed the green density gap between these powders but maintained
their relative positions. The density of $96\frac{1}{2}\%$ of theoretical obtained with sintered coarse iron hydro-pressed at 85,000 psi is
exceptionally high and warrants further attention.

The tensile and yield strengths and elongation are higher for
the fines, at the lower pressures, with the tensile strength rising
more rapidly than the yield strength. Above 60,000 psi the tensile
and yield strengths are about equal for the medium and fine powders while the elongation is slightly higher for the medium. Above
75,000 psi, the elongations drop off slightly while the strengths
continue to increase. The tensile strength of the coarse powder is
38% lower than for the fine at 60,000 psi, while the yield strength
is 16% lower and the elongation is 75% lower even though the
coarse material has the highest density. At least for the fine and
medium powders there appears to be a definite advantage for higher compacting pressures than 100,000. For the purpose of comparison to the maximum values of 38,000 psi tensile strength,
21,000 psi yield strength, and 15% elongation on a $\frac{1}{2}$-in. gage
length obtained on hydro-pressed iron powder, it is noted that

Table III. Comparison of Properties As-Sintered of Iron
Hydro-Pressed With and Without De-Airing

Property	Pressure	Fine		Medium		Coarse	
		De-aired	Not-de-aired	De-aired	Not-de-aired	De-aired	Not-de-aired
Sintered density,	25,000	75.5	70.8	80.0	77.3	—	—
% theoret.	60,000	85.5	83.7	90.0	89.5	92.9	93.3
Tensile strength,	25,000	19,440	17,800	13,000	14,900	—	—
psi	60,000	26,000	27,500	26,000	26,000	16,500	23,400
Yield strength,	25,000	10,320	9,500	6,600	8,800	—	—
psi	60,000	12,500	11,500	13,000	13,900	10,600	12,200
Elongation, % in	25,000	11.4	10.5	5.5	8.6	—	—
$^1/_2$ in.	60,000	16.0	22.7	18.0	20.7	4.6	21.2

wrought iron has the nominal properties of 42,000 to 52,000 psi
tensile strength, 26,000 to 35,000 psi yield strength, and a per
cent of elongation in 8 in. of 25 to 40%.

A comparison of the sintered tensile strength and green density (Fig. 15) shows that these properties rise at the same rate
above 40,000 psi. The higher green density of the medium powder
offers no advantage when compared to the fine for sintered strength.
Since its strength is as good as the fine above 40,000 psi, the medium should be used for structural parts, since it is somewhat
cheaper than fine powder.

In Table III, the sintered properties of de-aired and not-de-
aired hydro-pressed iron are compared. These results show that
the de-aired density is generally a little higher than the not-de-
aired material. At 25,000 psi, the strength and elongation are
slightly higher for the de-aired fine material while the reverse is
true for the medium powder. There is not any significant difference between the strength of de-aired and not-de-aired material
at 60,000 psi except with the coarse iron. In the latter case the
tensile strength was increased 42% by not de-airing and the yield
strength by 21%. The elongation was also increased in all cases
at 60,000 psi by not de-airing with a very large increase noted for
the coarse material.

If any generalizations can be made from these results it appears that with fine material slightly better properties are ob-

tained by de-airing the powder, with the medium powder any advantage is for not de-airing, whereas, with coarse powder there is a very definite advantage to not-de-airing. Indeed, the results with the not-de-aired coarse powder are very promising for its properties are close to those obtained with a fine and much more expensive powder and it shows an exceptionally high elongation for a powder product.

Tungsten

The sintered properties of the hydro-pressed tungsten (Fig. 16) were not always improved by increasing forming pressure. Unlike iron, the density of the tungsten fines reached a maximum at 15,000 psi and then dropped off slowly with increasing pressure. The medium and coarse, however, showed increases of 20 and 45%, respectively, over the pressure range. It was also found that while the fines produced the lowest green density of the three grades of tungsten, it yielded the highest sintered density. The order of the medium and coarse was unchanged by sintering; the medium is still denser than the coarse, but the gap between them has widened, since the coarse did not change much upon sintering.

The density of the fines was increased slightly by both granulation and by not de-airing, with granulation producing a greater improvement than not de-airing. The coarse material did not follow this trend; the density was unchanged by not de-airing.

The particle size of the tungsten has a tremendous effect on the sintered strength. The sintered fines, which were considerably stronger than the mediums or coarse, reached their maximum strength at a forming pressure of 25,000 psi, obtaining a value of 70,000 psi. Higher pressure caused a fall-off in strength. The mediums also had a similar maximum, peaking at 60,000 psi forming pressure with the much lower transverse strength of 46,000 psi. The coarse powder showed a steady increase of strength with pressure, but even at 100,000 psi only reached the low transverse strength of 14,500 psi.

Although granulation and not de-airing improved the sintered density, it has an opposite effect on strength. Granulation had the more deleterious effect of the two variables in that it reduced the strength of the ungranulated material by 42% while not de-airing produced a reduction of 21% at 60,000 psi and 19% at 25,000 psi, when compared to de-aired strengths.

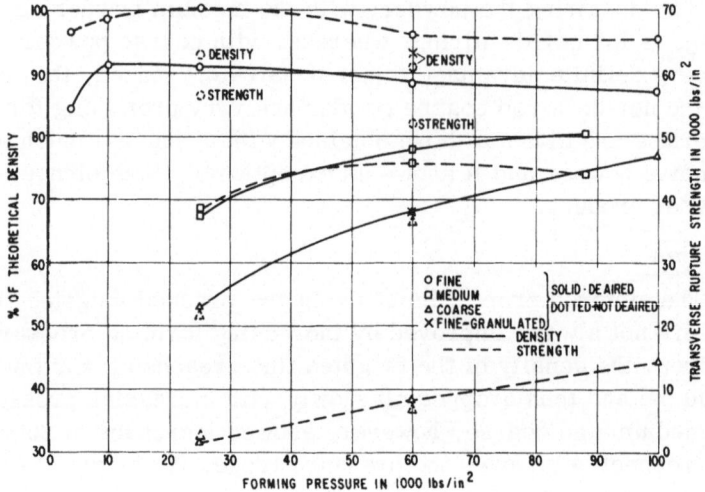

Fig. 16. Sintered density and sintered strength of hydro-pressed tungsten powder versus forming pressure.

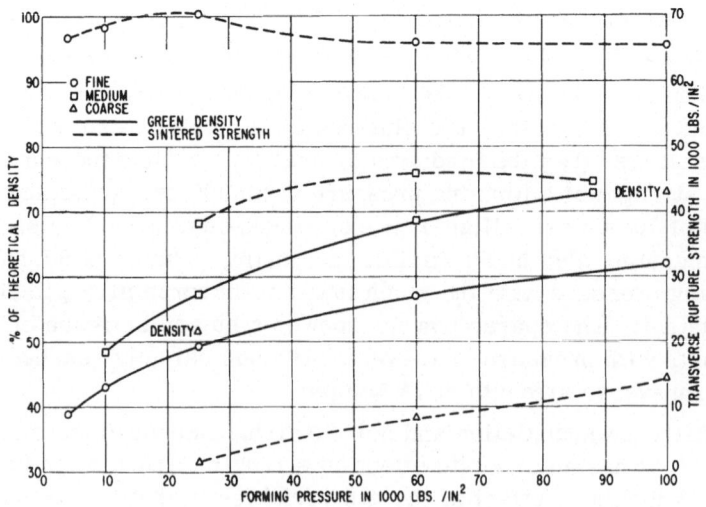

Fig. 17. Green density and sintered strength of hydro-pressed tungsten powder versus forming pressure.

A direct comparison of the green density and sintered strength of tungsten (Fig. 17) illustrates the point that high green densities are not always necessary or beneficial in obtaining high sintered properties. Although the green density of all three grades of tungsten rises with increasing forming pressure, only the coarse material shows a corresponding increase in strength. The medium and fine materials reach a maximum sintered strength at relatively low pressures and above these pressures the strength proceeds to fall even though the green density continues to rise. The particle size also has a tremendous effect on the sintered properties, for while the fines pressed to the lowest density, they sintered to the highest density and strength.

Comparing the results for tungsten and iron, some notable differences are observed. Iron showed no optimum forming pressure for maximum strength and density, but rather a continual improvement in these properties with pressure. The spread between fine and medium iron was relatively small, even though the particle size differences were greater than in the case of tungsten. Coarse iron gave the highest sintered density and when not-deaired, the strength approached those obtained with fine and medium iron. Coarse tungsten was far inferior to the other powders, and the effect of not de-airing was nil. The reason is not clear for the different behavior of these two metals, but certainly particle hardness must play an important role.

Alumina

The results obtained with alumina in some ways quite closely matched those exhibited by tungsten. The sintered density of fine alumina (Fig. 18) increased 3% as the forming pressure was increased from 4000 to 10,000 psi, and only 1.5% more when the pressure was further increased to 100,000 psi. The granulated density is very slightly higher than the ungranulated fines up to 10,000 psi and falls on the same curve above 10,000 psi. The coarse alumina's density, which is markedly lower than for the fine, increases with increasing pressure up to 60,000 psi, showing only a slight improvement above 60,000 psi. De-airing had no effect on the density of the coarse material.

The sintered strength also leveled off above 10,000 psi and in the case of the granulated fines appeared to show a maximum at

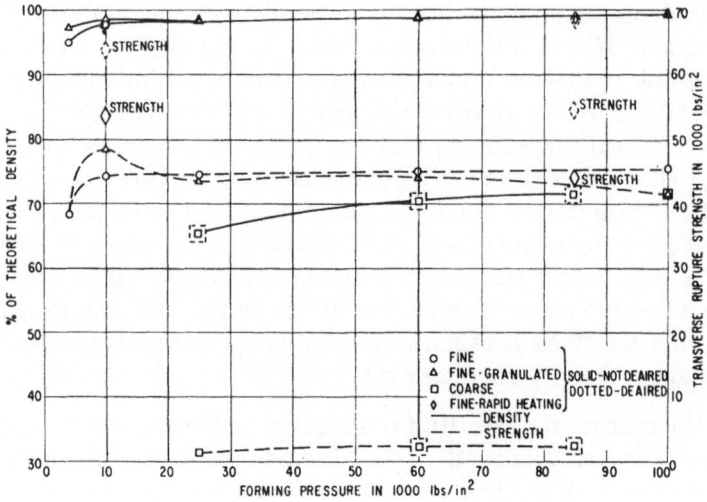

Fig. 18. Sintered density and sintered strength of hydro-pressed alumina
powder versus forming pressure.

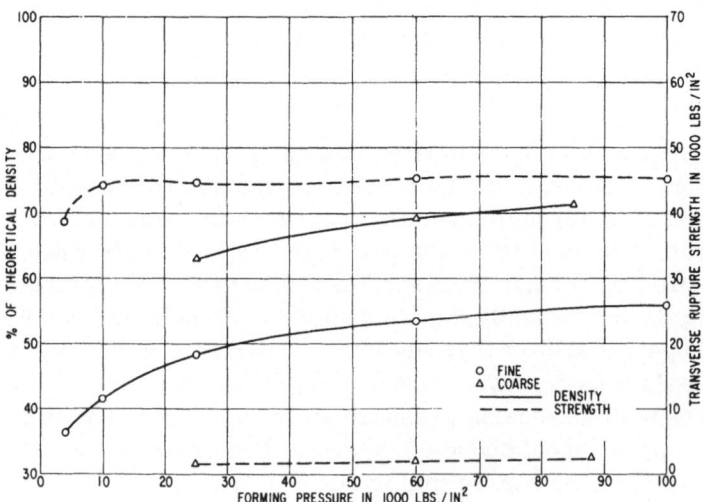

Fig. 19. Green density and sintered strength of hydro-pressed alumina
powder versus forming pressure.

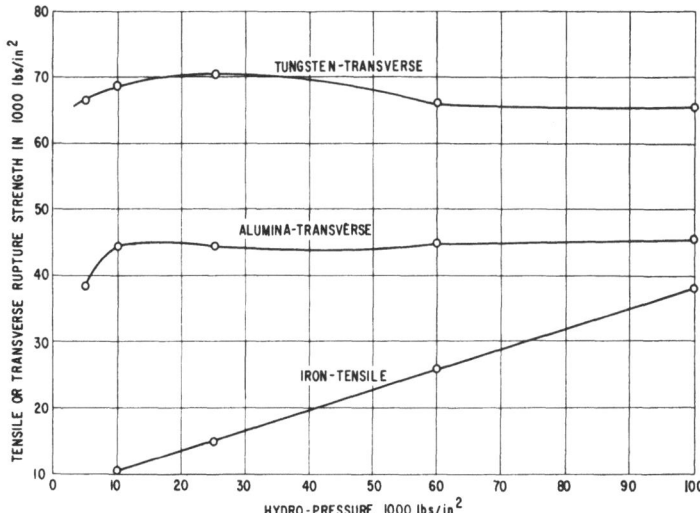

Fig. 20. Sintered strength of hydro-pressed iron, tungsten, and alumina powders versus forming pressure.

this forming pressure. Above 10,000 psi, the strength of the granulated material followed closely the ungranulated, gradually dropping away at the higher pressures. The large difference in green strength of these two materials apparently does not carry over into the sintered data. The strength of the coarse is very low over the whole pressure range, showing only a slight increase in strength with increasing pressure. De-airing versus not de-airing the coarse powder, as with tungsten, produced no change in strength values.

A series of fine powder alumina bars were processed with both de-aired and not-de-aired bags using a rapid heating rate for sintering. The sintering temperature was reached in 1 hr versus the 16-hr treatment described in Table II. Everything else was held constant. This treatment had no effect on the density, but did increase the strength significantly at 10,000 psi. The strength of the not-de-aired bars fast fired (solid triangle at 10,000 psi) is 23% higher than the same material slow fired. At 85,000 psi, this same treatment produced a slight drop-off. When de-aired, fast firing produced an even greater improvement at 10,000 psi (dotted triangle at 10,000 psi), some 45% increase. The strength dropped from 64,000 to 54,000 at 85,000 psi forming pressure, but this is still 20% better than the slow-fired samples. It is unfortunate that time did not permit the testing of de-aired slow-fired fine alumina, for it is likely that this would show that de-airing is beneficial for this material. Just why fast heating is so beneficial has not been determined, but it is possible that grain growth is suppressed by the shorter total time at elevated temperatures.

Figure 19 again shows that high green density does not guarantee the best sintered properties under a specific sintering condition. The green density of the fine is lower than that of the coarse, and both increase with the forming pressure, yet the sintered strength of the fines is not improved above 10,000 psi and practically no improvement is obtained with the coarse up to 88,000 psi. Indeed, the strength of the coarse is too low to be of any practical value.

Summary of the Hydro-Pressing Results

From the foregoing results, it is clear that the optimum hydro-pressure for producing maximum sintered properties varies with the hardness of the powder. This is graphically shown in Fig. 20, where the sintered strength of fine iron, tungsten, and alumina are plotted against forming pressure. The relatively soft or ductile iron shows a continual improvement with forming pressure whereas the harder tungsten and alumina both show a maximum at a low forming pressure, and then as the pressure is increased a gradual drop-off in strength occurs. It is interesting to note that at 10,000 psi alumina has reached its maximum strength, tungsten is nearly at its peak strength, but iron is extremely weak. This is especially surprising when one realizes that the iron was sintered at 78% of its absolute melting point, whereas tungsten was sintered at only 57% and alumina at 88%. It does appear that there is a different sintering mechanism operating for iron than for the two harder materials. The evidence is far from conclusive, for among the many differences that exist between the powders and their treatment, there is the very important fact that the fine iron is still considerably coarser than either the fine tungsten or alumina.

In Table IV, the authors have attempted to summarize the conditions that develop the optimum green and sintered properties of the eight powders studied. Some generalizations can be made from these results:

Table IV. Summary of Conditions for Maximizing Properties

Property	Material	Powder			Pressure—1000 lb/in.			De-aired	
		Fine	Med.	Coarse	10–25	60–100	>100	Yes	No
For Maximum Green Properties									
Density	Iron			×			×	×	×
	Tungsten		×	×			×	×	
	Alumina			×		×		×	×
Strength	Iron	×					×		×
	Tungsten	×					×	×	
	Alumina	×					×	?	?
For Maximum Sintered Properties									
Density	Iron			×		×			×
	Tungsten	×			×				×
	Alumina	×			×				×
Strength	Iron	×					×		×
	Tungsten	×			×			×	
	Alumina	×			×			?	?
	Alumina-fine-rapid heating	×						×	

1. For maximum green density use (a) coarse powders; (b) highest pressures; and (c) either de-airing or no de-airing with iron and alumina, de-airing with tungsten.

2. For maximum green strength use (a) fine powders; (b) highest pressures; and (c) not-de-aired iron, de-aired tungsten, and probably de-aired alumina.

3. For maximum sintered density use (a) coarse iron, fine tungsten, and alumina; (b) high pressure for iron, low pressure for tungsten and alumina; and (c) not-de-aired powders.

4. For maximum sintered strength use (a) fine powders; (b) highest pressure for iron, low pressure for tungsten, and alumina; (c) not-de-aired iron, de-aired tungsten, and probably alumina; and (d) fast heating rate for alumina.

Comparison of Die Pressing to Hydro-Pressing

The properties of die-pressed bars pressed in the rectangular steel die previously described were measured under the same conditions and with the same powders used for hydro-pressing in order to compare the two processes. The coarse metal powders at best were difficult to die press and at pressures below 60,000

Table V.   Comparison of Die Pressing and Hydro-Pressing as Shown by Green and Sintered Properties of Powder Iron Compacts

Material, pressure and processing method	Green density, theoret. %	Green strength, psi	Sintered density, theoret. %	Sintered strength, psi	Shrinkage, %* $\perp$	Shrinkage, %* $\parallel$	Elongation, % in 1/2 in. gage length
Fine Iron—25,000 psi							
Die pressed	63.4	1,700	68.9	13,900	1.3	2.8	6.8
Hydro-pressed—not-de-aired	68.3	1,400	70.8	17,800	1.5	1.8	10.5
Hydro-pressed—de-aired	70.0	1,750	75.5	19,400	—	—	11.4
Fine iron—60,000 psi							
Die pressed	78.1	5,000	81.8	25,300	1.0	3.0	14.7
Hydro-pressed—not-de-aired	82.5	5,600	83.7	27,500	1.0	1.6	22.7
Hydro-pressed—de-aired	81.0	4,600	85.5	26,000	—	—	16.0
Medium iron—25,000 psi							
Die pressed	70.7	—	73.9	8,100	0.5	2.1	5.1
Hydro-pressed—not-de-aired	75.5	550	77.3	14,900	0.5	0.5	8.6
Hydro-pressed—de-aired	75.5	650	86.0	13,000	—	—	5.5
Medium iron—60,000 psi							
Die pressed	83.1	—	90.0	21,200	0.3	1.5	10.4
Hydro-pressed—not-de-aired	88.5	2,600	89.5	26,000	0.24	0.28	20.7
Hydro-pressed—de-aired	85.2	2,000	90.0	26,000	—	—	18.0

* $\perp$: perpendicular to the pressing direction. The longitudinal direction for hydro-pressing. $\parallel$: parallel to the pressing direction.

psi were not strong enough to handle. The alumina powders also could not be die pressed with the existing equipment without the addition of a binder, which would defeat the intention of the comparison.

It is of importance to point out that die pressing is being presented in its most favorable light, since the size of the pressing assures good uniformity of properties. Thicker samples or variable cross sections would introduce density and strength variations not normally found in hydro-pressed bodies.

The green and sintered properties of die-pressed iron bars, as would be expected, followed the same pattern as the hydro-pressed bars.

The data are compiled in Table V along with the properties of the hydro-pressed iron both de-aired and not-de-aired. It can be readily seen that hydro-pressing iron powder produces higher green density and strength than die pressing, although the green strength of the die-pressed fine is fairly close to that of the hydro-pressed fine. The die pressings should be first compared to the not-de-aired hydro-pressings, since these conditions are more comparable. In the case of iron, this is the more favorable condition for hydro-pressing in most instances.

The sintered density and strength of the hydro-pressed iron is also higher than it is for die pressing. The strengths of the die-pressed bars are lower than both the de-aired and not-de-aired hydro-pressed strengths. The gap between the highest hydro-pressed value and the die-pressed strength has widened from what it was green, particularly at 25,000 psi, where there is a difference of 39% for the fines and 84% for the medium. At 60,000 psi, the hydro-pressed strength is 8.7% higher for the fines and 23% higher for the medium than the die-pressed values.

The shrinkage of the hydro-pressed bars is more uniform in the two directions than it is for die pressing. This results in better shape and tolerance control even though the per cent of shrinkage in the perpendicular direction is not reduced by hydro-pressing.

The most significant advantage of hydro-pressing iron powder is the improvement obtained on the elongation of sintered bars. At

Table VI. Comparison of Die Pressing and Hydro-Pressing as Shown by Green and Sintered Properties of Powder Tungsten Compacts

Material, pressure and processing method	Green density theoret. %	Green strength, psi	Sintered density theoret. %	Sintered transverse strength, psi	Shrinkage, %[a] ⊥	Shrinkage, %[a] ‖
Fine tungsten—25,000 psi						
Die pressed	50.7	—	92.3	69,200	20.4	22.7
Hydro-pressed—not-de-aired	48.8	900	93.0	57,000	19.9	19.6
Hydro-pressed—de-aired	48.5	735	91.0	70,500	18.0	17.7
Fine tungsten—60,000 psi						
Die pressed	56.7	—	92.8	64,500	17.5	16.9
Hydro-pressed—not-de-aired	57.0	1,800	91.5	52,000	15.2	15.0
Hydro-pressed—de-aired	57.0	2,280	88.5	66,000	13.4	13.3

[a] ⊥: perpendicular to the pressing direction. The longitudinal direction for hydro-pressing. ‖: parallel to the pressing direction.

25,000 psi the elongation was increased 68% for the fines and 72% for the medium iron over die-pressing results. With forming pressures of 60,000 psi, improvements of 54% on the fine iron and 99% on the medium iron were realized with hydro-pressing. It is thought that the higher strengths and elongations obtained with hydro-pressing are, in a large part, due to the more homogeneous structure that is produced by this technique.

The comparison of the properties of tungsten powder obtained by the two processes does not favor the hydro-press process, as it did in the case of iron powder. In general, there is little difference between die pressing and hydro-pressing (Table VI). Die pressing has a slight advantage for green density at 25,000 psi, while hydro-pressing is but a modicum better at 60,000 psi. The effect on the sintered density is just the reverse, and considering the sintered strength, hydro-pressing has a very slight advantage of 1.8% at 25,000 psi and 2.3% at 60,000 psi over die pressing.

Turning to the per cent of shrinkage on sintering, hydro-pressing has a somewhat greater advantage over die-pressing tungsten powder. Hydro-pressing results in lower shrinkage, which is practically equal in the two directions, while with die-pressing the direction perpendicular to the pressing direction has a lower shrinkage than the direction parallel to the pressing direction.

Density variations within die-pressed bodies of fine tungsten, fine iron, and medium iron increase in that order and could be an explanation of why the differences between die and hydro-pressing also increase in the same order.

As was explained in the last paragraph in the section entitled "Specimen Preparation," the powders were also pressed single acting in a $\frac{3}{4}$-in.-diameter lubricated steel die into $1\frac{1}{2}$-in.-long rods, and the densities of $\frac{1}{4}$-in. slices taken off top and bottom were measured.

The density variations are plotted in Fig. 21 and show the very interesting fact that the density variation from top to bottom increases as the particle size increases. The fine iron, which is coarser than the fine tungsten, shows a bigger variation – about $3\frac{1}{2}$ units to 2. The medium iron and coarse tungsten have about the same spread – 8 to 9 units. Also, it is of interest to note that

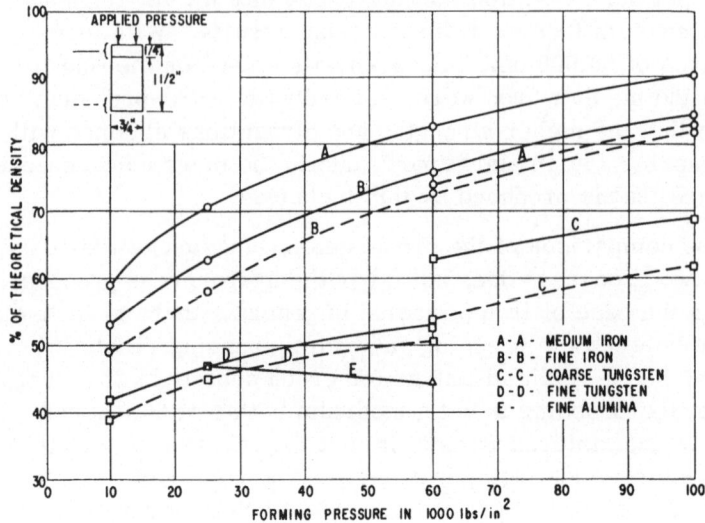

Fig. 21. Green density variations in round billets of iron, tungsten, and alumina prepared by die pressing.

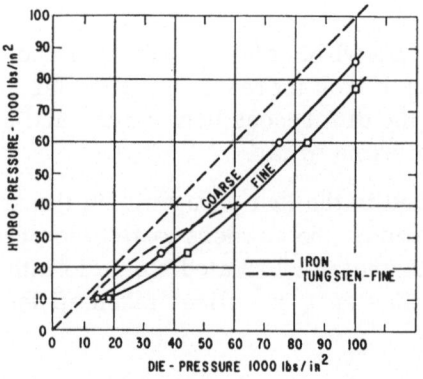

Fig. 22. Relationship between hydro-pressure and die pressure to produce equivalent pressed densities. Densities for die pressings were selected from the top $\frac{1}{4}$-in. section of the $\frac{3}{4}$-in.-diameter compacts.

the spread between the top and bottom densities for any given powder remains quite constant over the entire pressure range that
could be measured. The information above or below the abruptly
ending curves could not be obtained. Only the top segment of the
pressed alumina could be obtained, and this showed a decrease of
density with increasing pressure which is quite a different picture
than the green density curve of hydro-pressed alumina shown in
Fig. 12.

It has been reported by several sources [2] that very little
density variations exist in hydro-pressed bodies regardless of size.
Also, it has been stated that with a given powder and pressure, the
same overall density has been obtained with billets of from 1 to
10 in. in diameter by 12 in. long. Preliminary results by the
authors on hydro-pressed fine iron have not confirmed these
claims, therefore indicating a need for further corroboration. For
example, the maximum variation obtained on a $1\frac{1}{2}$-in.-diameter
hydro-pressed iron powder bar was 6.2% when de-aired and 2.2%
not de-aired at 10,000 psi as compared to 7.5% for die pressing.
At 85,000 psi, hydro-pressing yielded variations of 1.0% when de-
aired and 2.6% when not de-aired, while die pressing had a 3.6%
variation.

While these results favor the hydro-press process, they are
not as favorable as other investigators claim. It has also been
found that the coarse powders can be better handled by hydro-
pressing, and in some cases where it was impossible to die-press
them at low pressures it could be done by hydro-pressing. Even
with the fines, improvements could be seen with the hydro-press
process. This was especially true with the fine tungsten powder
which had a tendency to laminate when $1\frac{1}{2}$-in. lengths were die-
pressed in a $\frac{3}{4}$-in.-diameter die but were prepared crack-free by
hydro-pressing.

Another interesting fact when comparing these two compacting processes is that the forming pressure required to obtain a
given green density is less for hydro-pressing than it is for die
pressing (Fig. 22). This curve shows the die pressure required
to produce the density achieved by hydro-pressing at a given pressure. The maximum (top segment) densities shown for die pressing in the preceding figure are used for the comparison. If the
same die pressures were required, the dotted straight line would

result, so the added die pressure required is indicated by the shift to the right of this line.

The increment in pressure for die pressing is less for coarse iron than it is for the fine iron, and above 40,000 psi die pressure the gap between the two iron powders and the increment over hydro-pressing remains essentially constant, about 10, 14, and 24 psi, respectively. The difference in pressure for the two processes when pressing tungsten powder is less than it is for iron powders in the lower pressure range. At 25,000 psi there is a difference of 28% which increases to 46% at 40,000 psi, and above this pressure it is rapidly diverging from the equi-pressure curve. This increment in die pressure over hydro-pressure to produce the same density is due essentially to wall friction. The increment, of course, would change with a change in die cavity size. Therefore, the absolute values here are not as important as the realization of the value of eliminating wall friction in hydro-pressing.

Conclusions

There are a number of conclusions and generalizations that can be made on the basis of the current work. The following itemizes these succinctly.

Hydro-Pressing – As-Pressed Properties

1. The green density and green strength increase with forming pressure. (a) The slope of the density versus forming pressure curve decreases with increasing particle hardness. (b) The higher the apparent density of the powder, the higher the green density for a given pressure. Coarse tungsten did not obey this rule.

2. Fine powders produce stronger green compacts and generally show a more rapid increase in strength as the forming pressure is increased than the coarse materials.

3. De-airing or not de-airing had no significant effect on green properties. The strength of coarse iron and tungsten was somewhat improved by not de-airing, whereas de-airing helped the strength of coarse alumina.

4. Granulation does not improve green density, and in the very hard materials appears to reduce green strength.

Hydro-Pressing — Sintered Properties

1. The increase in sintered density with forming pressure becomes less pronounced as the hardness of the particles increases. (a) With iron, the higher apparent density powder (coarse particles) results in higher densities. The fines produce higher sintered densities with the harder materials. (b) With tungsten, the fines show a maximum density at 10,000 psi forming pressure.

2. Ductile powders show an increase in sintered strength with forming pressure. When the powders get very coarse, strength drops off.

3. The fine, harder particles show an early optimum for forming pressure beyond which strength falls off or does not improve. This appears to be in the range of 15,000 to 25,000 psi forming pressure. (a) The strength of coarse hard particles is considerably poorer than the fines, and generally continues to increase slowly with forming pressure.

4. Granulation appears to have a negligible effect on sintered properties.

5. The effects of de-airing versus not de-airing varies with the type of powder. (a) Not de-airing improved the properties of iron powder bodies as the powder became coarser. (b) With the fine harder particles, de-airing is to be preferred, for it appears to produce higher strengths without any appreciable sacrifice in density. (c) This is especially true when comparing rapidly heated alumina to slowly heated alumina.

Hydro-Pressing Versus Die Pressing

1. In general, a given powder processed under the same conditions will possess better mechanical properties when hydro-pressed than when die pressed. (a) The greatest improvement appears to be in the ductility of the sintered bodies.

2. Uniformity in density of pressed bodies is somewhat better for hydro-pressed material, resulting in more uniform shrinkage during sintering.

3. Hydro-pressing can consolidate powders that can be formed only with difficulty, if at all, by die pressing. This is particularly true for very coarse powders and fine, hard powders.

4. De-airing is easy to accomplish with hydro-pressing but difficult with die pressing.

References

1. H.W. Newkirk and R.J. Anicetti, Bull. Am. Ceram.Soc. 37:471 (1958).
2. Private communications.

Chapter 4. Explosive Compaction

Explosive Compacting of Metal Powders

Gerald Geltman

Isomet Corp.
Palisades Park, New Jersey

The technique of explosive pressing of metal powders is a logical outgrowth of the familiar technique of hydraulic pressure consolidation. Exposive charges have been used in ordnance work for years to study stress waves, fracturing, and plastic flow in materials. They have been used for tapping blast furnaces, mining, quarrying, construction, and destruction.

The application of high pressure to a study of materials is not new. Monroe [1] and J. and B. Hopkinson [2,3] are considered pioneers in the field of impulsive loading, having published works respectively in 1888, 1872, and 1905.

As early as 1909, Bridgman [4] developed what is now called the Bridgman Anvil: a device which permitted the achievement of extremely high pressures by essentially hydrostatic means. There has been considerable recent work by Hall [5] and others in which hydrostatic pressures higher than two million psi have been reported. As a result of these pressures, many interesting new materials have been created: Borazon, cubic boron nitride; Coesite,*

---

*Named for L. Coes, Jr. (Science, July 31, 1953, p. 131).

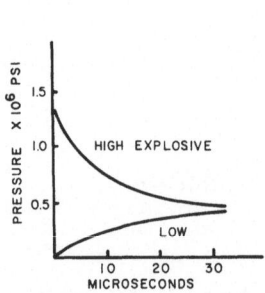

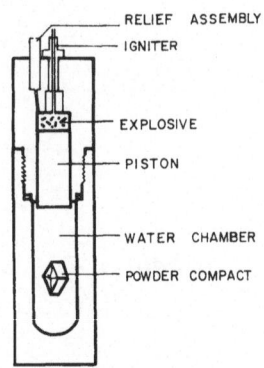

Fig. 1. Pressure—time curve for   Fig. 2. Simplified sketch of explosively
typical high and low explosive   activated hydraulic press (after McKenna,
(after references 6 and 13).   Redmond, and Smith).

a stable form of silica not attacked by hydrofluoric acid; cubic
tantalum nitride; and molybenum carbide. Some old ones — like
diamond — have been synthesized. Phase transformations have
been observed: bismuth with only one phase at atmospheric pressure exhibits seven other phases at pressures of 90,000 atmospheres.

The point of departure between high pressures developed hydrostatically and pressures developed explosively is, of course,
the rate of loading.

The variables of conventional powder pressing are relatively
well known, and numerous excellent treatises exist on the subject.
In the case of explosive compaction, several variables appear to
enter the picture for which often there are only speculative approaches to answers. There has been very little published data
available to the present on explosive compaction of powder metals.

The ordnance variables having an effect on compaction are:

1. Type explosive: low explosive such as smokeless
    powder and black powder, etc.; high explosive such as dy-
    namite, TNT, and RDX, etc. The low explosive is deflag-
    rating and characterized by a burning rate of hundreds of
    feet per second and pressures up to about 300,000 psi.

High explosive is detonating and is characterized by reaction zone travel velocities of several thousand feet per second and pressures of several million psi. Because a high speed of reaction in most high explosives results in a constant rate of energy release, regardless of the degree of confinement, high explosives are generally preferred. The control of the shock front is also more easily accomplished. Figure 1 shows the pressure–time curves typical of high and low explosives.

2. Quantity of explosive: The amount of explosive for a particular weight and shape of compact will, of course, have an important bearing on compaction and force generated. The explosive weight is actually expressed more conveniently as the ratio of metal powder to explosive. Therefore, lower numbers indicate an increase in explosive, and variations in explosive weight or powder can be readily correlated with other results.

3. Shape of explosive: The shape and the method of confining and positioning the explosive will determine the utilization of the developed pressure.

4. Duration and repetition of explosive force.

5. Relationship of charge to batch weight and part shape.

A treatise on "Behavior of Metals Under Impulsive Loads," by Rinehart and Pearson [6] provides valuable comprehensive background material on the concepts of impulsive loading.

There are many ways to utilize an explosive force to compact metal powders. The problem, however, is complicated by economic and safety considerations. Coupling of the generated force to the work piece may be accomplished in the following ways:

1. by shock waves traveling in a fluid, usually water or a medium which would behave in a manner similar to water, such as rubber;

2. by shock waves traveling in air;

3. by explosively forcing a piston into
   a. a chamber containing powder, or
   b. a fluid- or air-filled chamber where a bagged powder is suspended; or

4. by detonating an explosive which is intimately wrapped
    around the powder to be compacted.

One of the earliest reported utilizations of explosive in the
powder metallurgy industry was that cited by McKenna, Redmond,
and Smith [7]. A patent was obtained on an explosively activated
isostatic press used for compacting titanium carbide base cermets
and refractory metals. A piston was explosively forced into a
chamber containing a fluid and a rubber-bagged powder mix. The
advantages cited were high pressures, producing high density, uniform density, and elimination of subsequent warpage. In Fig. 2 is
a simplified drawing of the explosive-activated hydrostatic press
patented by McKenna et al. Such an arrangement could, of course,
be used with a shaped mandril. A tungsten carbide powder slug
compacted to a handling condition having a volume of 185 cubic
centimeters is finally compacted in the press to a volume of 130
cubic centimeters. In such compaction, 51,000 psi were developed. In general, pressures are deliberately limited to 50,000 to
60,000 psi for best compaction. The time for pressure development to produce best compacts had been found to be 25 to 50 msec,
typical of deflagrating low explosive.

The knowledge of the transmission of a shock wave through
water was applied to the compacting of metal powders by Montgomery and Thomas [8]. Powder was contained in a thin-walled
aluminum tube and immersed to a depth of 15 in. in water. Explosive placed above the work gave discouraging results. The
density of the compact was found to decrease with increasing distance from explosive. Subsequent tests utilized detonating fuse
(Cordtex) with a propagation rate of approximately 5000 ft/sec
wound around the aluminum tube. By this method solid as well as
annular cylinders were successfully produced. Titanium rods
were produced in lengths up to 20 in. and diameters as great as
$4\frac{1}{2}$ in. Their results are shown in Table I. It is interesting to
note that the powder bulk densities of the titanium, iron, and copper were, respectively, 33, 31, and 59% of the theoretical density. The iron at 31% required the highest relative weight of explosive expressed as the ratio of powder weight to explosive
(P/E). Although the importance of particle size is recognized by
the authors, no details are discussed. It would seem that the copper powders were fortuitously the most ideal mixture of particle

Table I. Compacting of Metal Powders* by Detonating Explosives
(Montgomery and Thomas)

Metal Powder	Powder (g/cc)	Weight of Metal Powder (g)	Powder Wt: Explosive Wt. [P/E]	Average Compact Diameter (cm.)	Average Compact g/cc	Density Percent Theoretical
W-titanium	1.53	675	11.2	3.5	4.0	89%
Aluminum	1.29	575	9.6	3.8	2.7	100%
Copper	5.25	2154	16.5	4.2	8.45	95%
Iron	2.5	1080	8.3	3.2	7.65	97%

*Container for powder: 17.5 cm × 5.5 cm diameter.
Metal powders: 100% passed 60 B. S. sieve.

sizes. The densities do not, of course, provide any indication of compact soundness or strength. A sample of lower density may in fact prove to be of higher strength than a high-density sample.

In compacting powders onto a mandril plastic, deformation of the mandril must be avoided — primarily to permit removal of the compact. The tensile strength of a thin-walled powder compact is likely to be less than the mandril and elastic deformation of the mandril would probably result in cracking of the compact. On the other hand, apart from problems of removal, compacting from the center of a female die outward to the die wall would increase the likelihood of compressive stresses on a cylindrical piece and improve the chances of compact integrity after removal of the die.

Cross [9] described simulated hot-explosive compaction. Lead powders were used since the lead recrystallization temperature is below room temperature. Rupture discs were used to limit pressure in a die 5 in. OD × 8 in. high with a $\frac{1}{4}$-in. ID. The upper end of the die was bored to receive a 0.25 caliber blank cartridge. The charge of low explosive was detonated. The shock front expanded in the confines of the cylinder and traveled through air to the work. The results are shown in Tables II and III. Increasing the pressure of compaction generally increased the compact compressive strength. In general, a peak in compressive strength was observed of compacts made from 40-micron powders. The very fine particle powders appear to require considerably more applied energy than coarse particles agreeing with the general experience of conventional powder pressing. Of course, die fric-

Table II. Effect of Oxide Film on Compressive Strength (Cross)

Powder Type	Ave. Particle Dia. Oxide Film Thickness [D/T]	Compressive Strength (psi) at Press Pressures of 20,000 and 50,000 psi	
SF	16.6	—	1900
40 B	26.2	3100	4000
F	52	4000	—
200 B	72	5200	6600

Table III. Effect of Particle Size and Explosion Pressure on Compressive Strength (Cross)

	Average Particle Size, Microns	Average Oxygen Content Weight Percent	Apparent Density g/cc	Compressive Strength(psi) Explosion Pressure psi		
				10,000	20,000	50,000
40B	131	0.59	5.67	2600	3100	3900
100B	59	0.14	6.10	2500	3200	2900
200B	41	0.17	5.70	2500	5300	6800
F	26	0.27	4.53	4000	4000	6400
SF	5	0.36	4.4	100	100	1800

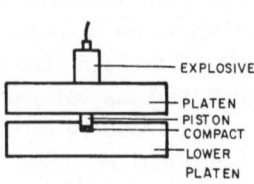

Fig. 3. Single piston press (after LaRocca and Pearson).

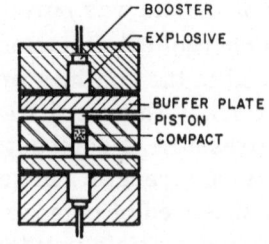

Fig. 4. Double piston press (after LaRocca and Pearson).

tion and physical travel of the powder should not be ignored as factors. The compressive strengths increase dramatically at 50,000 psi. But Cross has shown that the oxide film on the powders appears to play a critical role in compact strength. As seen in Table II, the highest average particle diameter to oxide film thickness gave the highest compressive strengths. It would be interesting to observe the results of a similar series of compactions with essentially oxide-free powders. It would also be desirable to avoid explosion gas contact with the powder, or at least evaluate the effects of such contact.

LaRocca and Pearson [10] describe the use of single- and double-acting presses (see Figs. 3 and 4) in which the pistons were activated by explosives. The dies are usually made of 1% carbon drill rod or low carbon steel.

As a result of dynamic compression, part of the beta phase of cobalt was transformed to the alpha, or hexagonal modification which is the more stable form of cobalt. The transition was determined by powder x-ray diffraction. Two factors were considered important in the double piston system: (1) both explosive charges must be detonated simultaneously, and (2) the explosive charges must be placed accurately with respect to the piston centers. It was further stated that some ductility in the test material is necessary if the material is to cohere under press action. Diamond dust and some ceramic compositions, which are not ductile, did not cohere and remained a powder. Brittle materials, unable to deform plastically, would most likely be reduced in particle size; an excellent, but expensive method for comminution. The final density was shown to be closely related to the ductility of the material pressed. It was indicated that thermal effects were only superficial or localized. Titanium powder was pressed to a density 95% of the theoretical.

In the single-piston press shown in Fig. 3, sectioning of the press was required to remove the specimen and is not considered practical except for research purposes.

In cases where air alone was used as a coupling medium, insufficient pressure was developed to produce significant compactions. In other experiments tubes were coated with explosive, but large tensile stresses that were set up caused the work and tool

to fly apart.   Pearson and LaRocca produced titanium to 95% of
theoretical as well as shaped compacts — crosses and washers.
They also compacted iron oxide with a small amount of aluminum
to a density of 80% of theoretical.

The addition of various metal binders or the use of metal-
coated brittle grains would improve the compactability of brittle
materials. However, excessive explosive and/or too soft a ductile phase would serve to transmit the shock to the dispersed
grains and fracture them.   Then, too, the degree to which a ductile additive could be tolerated metallurgically would have a bearing on its use.

The shape of the explosive and the sample have an important
bearing on compact integrity. A cylindrical charge has an essentially parabolic pressure front; analogous in many respects to the
pressure distribution or density gradient due to friction in a conventional die. It can be shown that solid discs might be fragmented
due to overlapping of shock fronts in a double-piston press,
while donut-shaped samples may be compacted to very high densities.

The ratio of height or thickness to diameter appears to be as
important in explosive pressing as it is in conventional pressing.

It is well known that the speed of pressure application greatly
influences the achieved density, green strength, and electrical resistivity of compacted metal powders. In conventional pressing,
rapid speed generally results in poor compacts. Speeds up to 5
ft/sec are the norm, while in explosive compacting, speeds in the
ten thousandths of feet per second are utilized with excellent results. By virtue of the rapidity with which the process takes place
it is impossible to distinguish the various stages of powder compaction. It is likely that in explosive compaction a greater degree
of cold welding and plastic deformation takes place than in conventional pressing.

Using a liquid medium for the transmission of the explosive
force has two important advantages:

1. longer pressure hold times, and
2. protection to the operators and/or facility.

It is for the second reason that the Dynapak process [11] is extremely interesting. The method employs gas pressure imbalance. The high-velocity press is operated on a cylinder of compressed gas. The development of compacting pressure depends on the mass of the piston and the force behind it. Using explosives permits high pressure and high velocity with lightweight pistons and a much cheaper investment in working tools. The Dynapak machine is easy to operate and lends itself to production cycles more readily than explosive compacting.

A brief report [12] of efforts to compact brittle materials in a Dynapak machine indicated largely a lack of success. Heating of the materials (glass, etc.) was said to improve the possibility of forming, but then hot-pressing, in conventional presses over a longer sinter-time permits solid state bond development, which is not possible in impulsive loading.

Other methods are constantly being developed utilizing chemical explosions of all kinds and electric discharge.

Explosive compaction of metal powders offers an exciting and unique tool for the production of high-density green compacts. For unusual materials and/or short runs, the method appears highly competitive. Expensive hydraulic equipment can be eliminated and, usually, less expensive dies may be employed.

References

1. C. E. Monroe, "Modern Explosives," Scribner's Magazine, 3 : 563-576 (1888).
2. J. Hopkinson, "On the Rupture of Iron Wire by a Blow," Proceedings of the Manchester Literary and Philosophical Society, 11 : 40 (1872).
3. B. Hopkinson, "The Effects of Momentary Stresses in Metals," Proc. Roy. Soc. (London), Vol. 74 (1905).
4. P. W. Bridgman, The Physics of High Pressure, G. Bell and Sons, London (1952).
5. H. T. Hall, "Ultrahigh Pressures," Scientific American, November (1959).
6. Rinehart and Pearson, Behavior of Metals Under Impulsive Loads, American Society for Metals, Cleveland (1954).

7. McKenna, Redmond, and Smith, U.S. Patent 2,648,125 (August 11,1953); filed (August 6,1947).

8. Montgomery and Thomas, "The Compacting of Metal Powders by Explosives," Powder Met., No. 6 (1960).

9. A. Cross, "Try Hot-Explosive Compacting for Sintered Powder Products," The Iron Age, December 24, 1959.

10. E. LaRocca and J. Pearson, "Explosive Press for Use in Impulsive Loading Studies," Rev. Sci. Instr., October 1958.

11. F. Monahan, "Precision Forgings and Extrusions by the Dynapak Process," Machinery, June, 1960.

12. Anonymous, "Impact Formed Glass and Ceramics Studied," Materials in Design Engineering, February 1962, p. 118.

13. C.S. Simons, "Explosive Metalworking," DMIC Memorandum 71, November 3, 1960.

Consolidating Metal Powders Magnetically

Donald J. Sandstrom

Staff Member, University of California
Los Alamos Scientific Laboratory
Los Alamos, New Mexico

The magnetic field produced in a coaxial conductor composed of two tubes carrying high pulsed currents can be used to compact powders of refractory metals and ceramic–metal composites into tubes over 30 in. in length. Densities approaching 100% of theoretical are reached with peak pressures above 50,000 psi.

We have devised a technique for consolidating refractory metal powders into tubular shapes by using the opposing magnetic fields in a coaxial conductor. With our device, we have compacted tubes of tungsten, molybdenum, $W + UO_2$, and $Mo + UO_2$ powders without need for binders in the powders. (Furthermore, it is possible to compact extremely long lengths at one time with a very uniform pulse and produce intricate shapes by varying mandrel geometry.) In every instance, green density and strength of the as-compacted tubes was high enough to allow us to handle them with ease even before sintering.

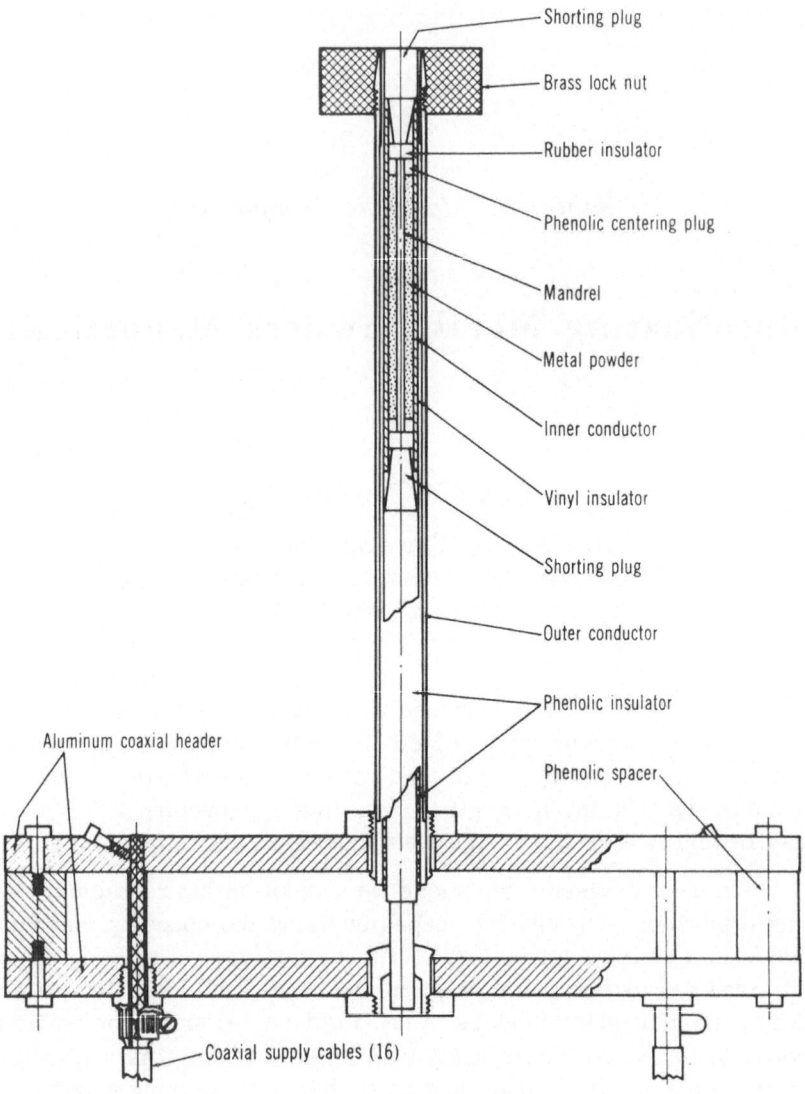

Fig. 1. Heavy currents through the coaxial conductors produce strong magnetic fields which act to crush metal powders into coherent tubes. When the mandrel is extracted, the copper shielding is etched away and the tube is sintered.

Table I. Density of Sintered Refractory Metal Tubing

Material*	Particle Size	Sintered†			Resintered‡		
		Open Porosity	Closed Porosity	Theoretical Density	Open Porosity	Closed Porosity	Theoretical Density
Molybdenum	4.5 micron	0.69%	5.57%	93.74%	0.27%	4.52%	95.21%
Molybdenum	1.8 ' micron	0.52	2.22	97.62	—	—	—
Mo+40 vol % UO₂	UO₂ = −250 +325 mesh						
	Mo = 1.82 micron	8.95	2.67	88.38	2.98	5.12	91.90
W+40 vol % UO₂	UO₂ = 4.2 micron	18.39	0.12	81.43	11.56	2.56	85.88
	W = 2.1 micron						
Tungsten	3.76 micron	2.49	10.31	87.20	—	—	—
Al₂O₃	Graded particle size	0.45	7.38	92.17	—	—	—

*Compacted over a steel mandrel.
† At 3100°F (1700°C) in hydrogen for 3 hr.
‡ At 3600°F (2000°C) in vacuum for 2 hr.

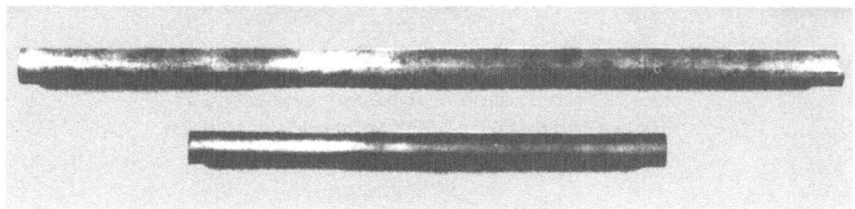

Fig. 2. Tubing of molybdenum powder can be magnetically compacted (top) and sintered (bottom) to nearly 100% of theoretical density.

How the Method Works

This device makes use of an interesting phenomenon – the ability of a conductor carrying very high pulsed currents (in the range of 10^5 to 10^6 A) to crush itself with the magnetic field generated around it by the current. By confining this magnetic field in an annular gap between the inner and outer sections of a coaxial conductor, it can be made to act as if it were a compressed gas, crushing the inner conductor or expanding the outer conductor.

Figure 1 depicts such a coaxial device in a setup used for consolidating metal particles into a tube. The outer conductor is strengthened with a sleeve of austenitic stainless steel; the inner conductor is weaker, being simply an annealed copper tube which acts to compress and confine the powders being compacted.

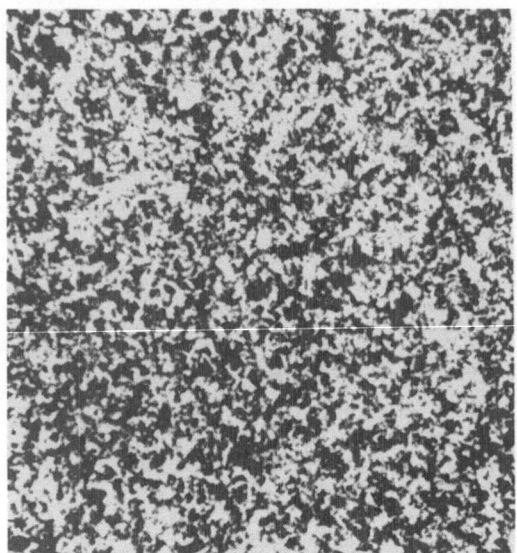

Fig. 3. As magnetically compacted, the microstructure of a tube made of molybdenum powder (4.5 μ)
displays little porosity. When sintered, the density is
approximately 95% of the theoretical value. 100 ×.

Tapered plugs, also of copper, are driven into the center tube to
produce good electrical contact at the tube ends. To prevent arc-
ing between the tubes, the gap between inner and outer conductors
must be insulated with a thin dielectric material; vinyl tubing is
effective, we have found.

The powder to be compacted is placed in the annular gap between the inner conductor and the mandrel. Tight-fitting rubber
plugs are inserted over the mandrel holders to seal the ends and
insulate the shorting plugs from the mandrel. The coaxial coil is
then firmly attached to an aluminum coaxial header (powered by a
bank of capacitors) by means of collets which grip the outer and
center conductors. When the capacitor bank is discharged, the
current is carried up the inside of the outer conductor and back
to ground through the inner conductor, generating magnetic fields
which repel each other. Because the inner conductor is weaker
than the outer conductor, it is crushed by the magnetic field.

The bank used to supply our coaxial device has a capacitance of 715 μF at a peak voltage of 20,000; it produces a peak current of 59.8 A/V and develops instantaneous peak pressures over 50,000 psi when compressing the powder.

Packing the Powder

Though the powder does not need to be packed densely, it requires uniform density to assure uniformity in the compacted tube. To fill the inner conducting tube, the mandrel is first centered in a thin-walled tube of cellulose acetate. (This plastic tube protects the powders from acid attack when the compressed tube of copper is dissolved after compaction.) The loose powder is then packed into the confinement sack on an ultrasonically vibrated table. Steel mandrels are generally preferred because fusible alloy mandrels, which may be used, make it difficult to keep the mandrel centered so that the compacted tube will be concentric.

Immediately after forming, end plugs and vinyl insulation are removed, and the mandrel is extracted manually or on a draw-bench. Though removal of long mandrels from compacted tubes can present a problem, we have circumvented it by the following technique. Before assembly, the mandrel is first lubricated with a thin film of silicone grease and slipped inside a polyethylene or Teflon "shrink" tube. (This tubing is a commercially available item which shrinks upon heating.) The plastic tubing prevents lubricant from contacting the compacted metal powders, while the lubricant itself allows the mandrel to be extracted rather easily. Later, the shrink tubing is removed by heating at 300°F (150°C) and the copper tube is etched off with acid. The inner conductor of copper should not be removed until the mandrel has been extracted from the compact, because it gives the compact enough mechanical strength to prevent compression cracks from developing in the green tube.

Sintering the Tubes

Compressed tubes are sintered in a resistance heated furnace containing hydrogen. Because the furnace's uniform heat zone is approximately 18 in., we have not been able to sinter tubes of greater length. However, we have compacted tubes up to 30 in. long, and longer lengths are readily attainable.

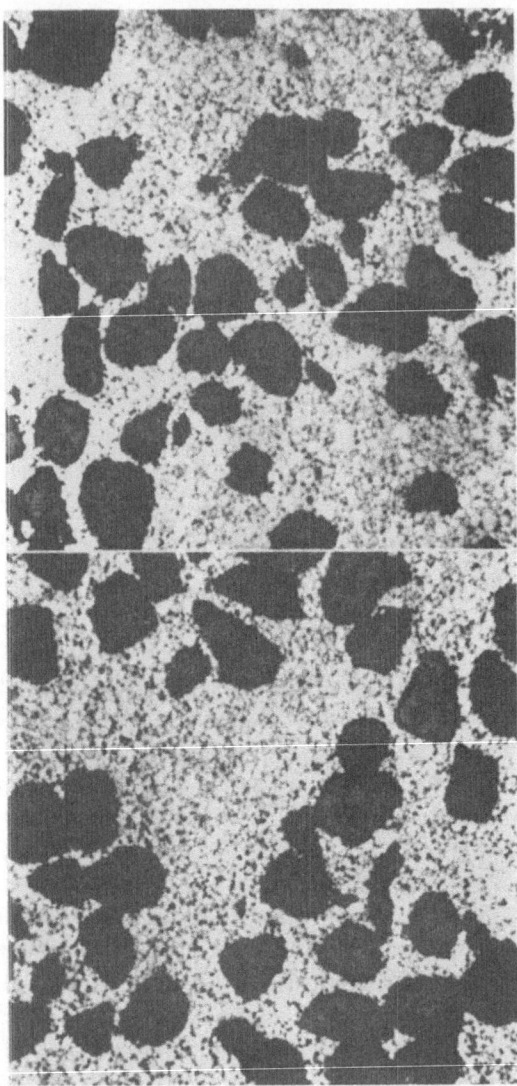

Fig. 4. Transverse(top) and longitudinal specimens of Mo + 40 vol.% UO_2 tubing, as compacted and sintered, appear exactly the same, indicating lack of directionality. 100 ×.

Tubes are sintered at 3100°F (1700°C) for 3 hr; in some instances, they have also been sintered in vacuum at 3600°F (2000° C) for 2 hr after the hydrogen sintering.

Figure 3 illustrates the structure of as-pressed molybdenum tubing, and Fig. 4 depicts photomicrographs of Mo + 40 vol.% UO_2 in longitudinal and transverse directions. Because the UO_2 particles have approximately the same shrinkage as the molybdenum matrix, the density of the sintered tubes was approximately 92% of theoretical after sintering in vacuum.

We also produced another type of pressed and sintered composite of tungsten with uranium dioxide in smaller particles: 4.2 μ. Because the shrinkage of this dioxide did not match that of the metal matrix, however, the cermet had lower density (approximately 86% of theoretical after vacuum sintering). Table I lists some of our results.

Something about the Future

To date, our work on the magnetic compaction process for fabricating tubes has been primarily a feasibility study. Now we are making a more systematic examination, evaluating, among other items, the effect of variation in peak pressure on green and sintered density. (In fact, we have noted that metal powders can be "overpressed," producing cracks which are analogous to cracks produced in cold pressing.) Also, the effects of particle size, particle size distribution, and variations in mandrel material are under study. Furthermore, a limited amount of work on the use of epoxies as binders indicates that they are excellent from the standpoint of low residual impurities upon "burnout" and high strength in the green cured state.

In conclusion, we have found that the magnetic technique of powder compaction is highly reproducible, safe for the operator (if the safety precautions for high-voltage, high-current systems are observed), and practical for operation in confined areas. The future should see significant advances in the use of this technique.

Chapter 6. Continuous Compaction

A New Method for Compacting Metal or Ceramic Powders into Continuous Sections

Frank Emley

Group Manager, Powder Metallurgy Group

and

Charles Deibel

Powder Metallurgy Group
Westinghouse Electric Corp.
East Pittsburgh, Pennsylvania

In the powder metallurgy and ceramic technologies, the largest field of application involves forming of powder raw material into the shape of the finished product by a molding or compacting technique, and the simultaneous or subsequent sintering of these powders into a strong aggregate by the application of heat.

Another field of application, the consolidation of powders into relatively large masses which may serve as end products themselves or as intermediates for further processing to final shape, is becoming increasingly important as a means of achieving many of the complex materials requirements of modern science and industry. The processing of tungsten filaments from powder raw material is a well-known example which derives its importance from the fact that tungsten, because of its high melting point, cannot practicably be fabricated into solid metal by any other process. Some of the unique advantages which may be gained in using powder raw materials to fabricate relatively large masses may be listed as follows:

1. Incompatible materials, e.g., ceramic and metal powders, can be intimately blended together and consolidated into a dense, solid mass without macro-segregation.
2. Many of the powder raw materials (both metal and nonmetal) are commercially available in a relatively high purity state. Furthermore, these same materials are usually very consistent in impurity content. In the case of metals, especially, many of the impurities are removed during the sintering operation. For these reasons, and since powders can be mixed with precision, solid masses may be made from powders to very precise composition limits and with low and uniform impurity levels.
3. Many alloys are not amenable to commercial production because they exhibit extreme brittleness as large cast ingots. Metals and alloys of this type, such as beryllium and 6.5% silicon–iron, have been fabricated from sintered powder billets by conventional processing techniques of forging and rolling.
4. Materials having certain properties, such as controlled porosity, can be produced only from powders.
5. Metals with exceptional strength at elevated temperatures resulting from a controlled dispersion of an insoluble phase have thus far been produced successfully only from powders.

These advantages to be gained in the forming of large sections, especially in the field of powder metallurgy, have not been fully exploited, because of limitations of current fabricating techniques. Some of these limitations are pointed out in the following paragraphs.

Powders are most commonly compacted in closed dies using relatively high pressing pressures. Even the highest capacity presses will produce shapes of very limited volume by this method.

Fig. 1. 100-lb. ingots of iron–nickel alloy pressed
hydrostatically.

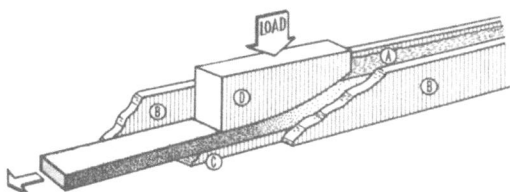

Fig. 2. Cutaway schematic view of continuous compaction setup.

Larger amounts of powder can be compacted hydrostatically
in rubber containers. Although a practical maximum size of several hundred pounds can be achieved in this manner, expensive
equipment and a cumbersome technique restrict its application.
As a matter of interest, the photograph in Fig. 1 shows two 100-
lb. ingots of a magnetic alloy which were pressed hydrostatically.

Extrusion of metal or ceramic powders, usually combined
with large percentages of binder, can produce items of considerable but definite volume. This process has been confined to very
specialized applications.

Gravity sintering has been used experimentally to consolidate
large masses of powder. In this method, powders are packed
loosely in a container with gravity being the only compacting force.
In order to obtain relatively high densities by this method, a great
deal of shrinkage must be tolerated. Lengths over a few inches
tend to tear apart during this extensive shrinkage.

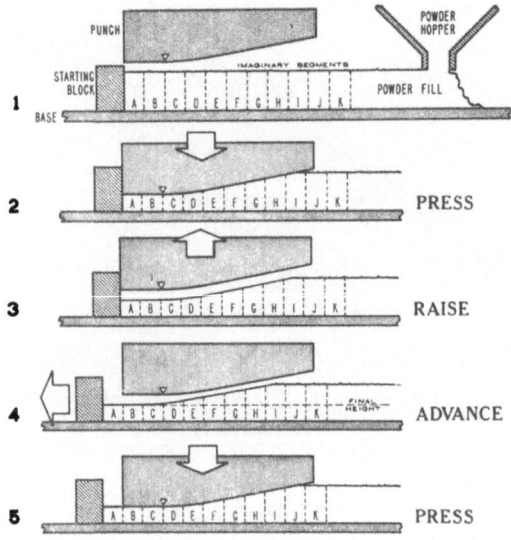

Fig. 3. Sequence of operations in continuous compaction.

Fig. 4. View of bar surface during pressing.

Slip casting has been used for many years in the ceramic industry and has just recently been successfully developed as a process for the forming of metal powders into complex shapes. By this method it is also possible to achieve relatively large masses; nevertheless, practical limitations are obvious.

Direct rolling of powders is the only important method being exploited at the present time for the continuous formation of solid materials from powders. The method entails the use of rolling mills similar to those used in conventional metal rolling. Powder is fed into the rolls, where it is compacted into green sheet which may be subsequently or continuously sintered and then rerolled to high density. In this process, it is well established that the maximum thickness of sheet is primarily dependent on roll diameter. All reports which have come to the attention of the authors indicate that even with rolls of very large diameter, the upper limit of thickness for green sheet or strip appears to be about $\frac{1}{4}$ in. Conventional rolls will produce green strip to a maximum thickness of only $\frac{1}{16}$ in. or less. Roll speed, another important variable in this process, must be kept low.

It is thus apparent that a very desirable innovation to the art of powder metallurgy and to allied powder arts would be a pressing technique capable of producing bars of unlimited length and relatively large cross-sectional area. We wish to describe such a technique for powder pressing which has been developed at the Westinghouse Materials Engineering laboratories and to discuss some results which we have achieved using our method. Patent applications covering this process have been filed.

The method, which we call continuous compaction, can best be described by illustrating a typical setup for laboratory studies. Figure 2 is a cutaway sketch of such a setup. Loose powder marked A is placed into a three-sided trough with sides marked B, and a bottom marked C. The powder is compressed with a punch marked D, which consists of a flat pressing surface parallel to the powder surface and also a pressing surface which is angled to the powder surface. These two surfaces are faired together.

Figure 3 illustrates the sequence of events during pressing and will essentially describe the major features of the process.

The punch is raised clear of the pressing area in step one and loose powder is charged into the channel formed by the die walls, which are not shown. A starting block is used to contain the loose powder fill until pressing is under way. For convenience in illustration, the loose powder fill is shown divided into a series of imaginary segments labeled A through K.

In step two, the punch is lowered into the channel. The horizontal flat surface, which will be called the "finishing area," is adjacent to the starting block. When sufficient pressure is applied to achieve the desired degree of compaction, a study of this picture will show that, over the length of the sloping portion of the punch which is in contact with the powder, the powder has been pressed with an infinite number of pressure values from zero in a portion of segment I to the maximum pressure available under the finishing area in segments A and B.

Step three of this illustration shows the punch raised after the pressure stroke. The distance raised is arbitrary except that a minimum travel is necessary in order to avoid interference of the punch with the powders during the next step.

In step four, the powder is moved as a unit a prescribed distance to the left. The significance of this step is obvious from the illustration. All the powder that was under the finishing area of the punch in step 2 has been subjected to maximum pressing pressure in step 2 and, consequently, has been compacted to its desired green density. If the powders are then moved to the left a distance less than the length of the "finishing area" of the punch, some of the fully compacted powder, in this case segment B, will still remain under the "finishing area." Since it cannot be compacted any greater amount, assuming the same or slightly less pressure is again applied, it will show no density discrepancies nor surface marks from the left or exit edge of the punch. All the powder now under the punch, however, except segment B, is in position to receive another increment of compaction during the next pressure stroke.

Step five is identical to step two, except that the punch is now compacting the powder fill slightly to the right of the previous stroke. Segment C, being under the "finishing area" of the punch, has now received full compaction. Segments D through I have received additional increments of compaction, and a portion of segment J has received an initial increment of compaction. By cyclic repetition of steps 3, 4, and 5, the loose powder is pressed into a continuous bar.

The method which has been described thus far in general terms can utilize an automatic setup capable of producing bars of unlimited length that can be fed directly into a sintering furnace and thence to further processing steps such as rerolling. A mechanism can be readily designed for feeding the powder onto a carrier which automatically moves it to the pressing mechanism and transports the pressed bar into the next processing stage. For experimental work with the process, however, a much simpler setup has been designed and built which can produce bars of limited length. The punch is attached to the upper ram of a manually controlled hydraulic press. Instead of an automatic powder feeding and transporting mechanism, a simple U-shaped channel has been constructed with removable sides for easy removal of the pressed bar. Figure 4 is a photograph of an early laboratory setup which shows the punch raised and gives a good view of the powders inside the channel about half way through the operation. In this setup, the cyclic operations of pressing and advancing the powders are manipulated by hand in accordance with the steps shown in Fig. 3. The powder and pressed bar are advanced by moving the entire channel forward. When the right end of the channel is reached, the pressing operation is stopped and the green compact removed by dismantling the channel.

Experiments have been conducted to determine the general characteristics of this pressing process and to study the effects of some of the variables on the pressed bar. The effect of load on green and sintered densities, the effect of punch configuration, the effect of varying the forward movement per stroke, and the effect of varying the powder fill are some of the details of the process which have been studied. In addition to this, a large number of different powders and powder mixtures have been pressed in this apparatus. Results of these investigations are presented in a qualitative manner in this paper.

Certain features of the process should be pointed out in more detail. The slope of the punch is not critical, since experiments have shown that it may vary over a wide range of angles from the

horizontal. Low angles reduce the speed and efficiency of the process but may be more desirable for pressing a small powder fill;
high angles cause excessive quantities of powders to be laterally
displaced, resulting in bar thickness variations. A proper balance
must be established depending upon the end requirements. The
sloping face of the punch need not be a smooth surface as shown
in the illustrations, but may have, for example steps as shown in
Fig. 5. Such a punch increases the angle at which pressing can be
accomplished to the point where it is limited only by the magnitude
of the pressure gradient which a given powder can sustain.

It is interesting to compare this method of compaction and
closed die compaction. In closed die compaction, the powder is
contained in all directions by the four walls of the die (assuming a
rectangular die) and the two punch faces. In the continuous compaction process, however, the compressed bar serves in place of
one of the die walls to restrain the powder in one direction and the
gentle slope of partially compressed powder restrains the powder
in the other direction. The two side walls of the trough or die
must, of course, be at least as high as the desired powder fill.

Many will recognize that starting the cyclic operation intro-
duces a boundary condition not again encountered in the pressing
of a bar. This condition can readily be overcome, however, by using a very compressible material such as sponge rubber in the initial portion of the fill, so that the punch only meets the same compacting resistance in the first stroke as it does in subsequent
strokes. Actually, however, very little difficulty is encountered
even if this measure is not carried out.

A wide variety of cross-section shapes and sizes can be obtained in our continuous compaction process. We have used 1-in.-
and 2-in.-wide channels in our experimental work, but there is no
reason why wider bars cannot be pressed. The required capacity
of press, other conditions being equal, will vary directly with
width. Thickness is not severely limited, although a practical
limit will probably be encountered in any given press. The authors
have achieved pressed bar thickness up to 1 in. with no difficulty
and have not attempted any thicker pressings. The lower limit of
thickness has not yet been established, but it is assumed by the
authors that materials could be compacted into strip of 5 to 10
thousandths of an inch thickness with no difficulty. As noted pre-

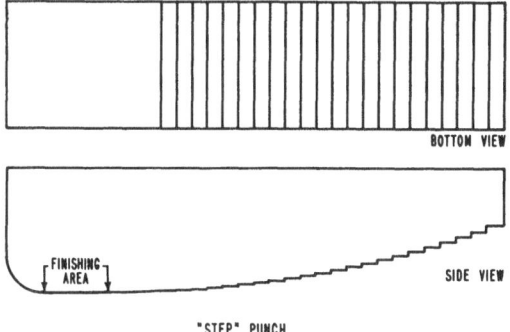

Fig. 5. Sketch of an alternate punch design.

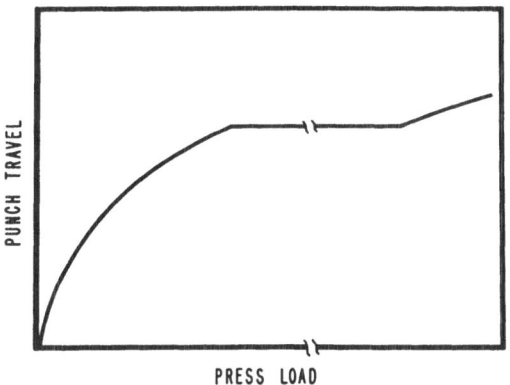

Fig. 6. Theoretical curve of punch travel versus press load.

viously, any length is possible and even practicable. Pressing of
fixed length bars of any reasonable length can be achieved with
relatively simple equipment as compared with complex equipment
required for truly continuous operation. Even a fixed length pressing setup, however, can be operated automatically.

Section shapes can achieve the entire range now possible with
conventional powder metallurgy pressing if one considers those
shapes permissible in sections parallel to the pressing direction
with so-called "single action" pressing. Multiple pressing actions
can be envisioned, at least from the top pressing surface.

Bars produced by this method are constant in cross section over the entire length if the powder fill entering the pressing area is maintained constant. Under this condition, the top or pressed surface will be smooth and regular and, of course, parallel to the bottom surface. Another important advantage is the consistency of green density which can be achieved readily over the entire length of the pressed bar. The excellent quality of the surfaces, edges, and corners of the finished bars made by this process is an important feature.

The good control of density and thickness tolerances noted above is understandable when one considers the variation of press load with punch travel (or green density) during the pressing stroke. The schematic curve shown in Fig. 6 (derived from green density versus pressing pressure considerations) has increasing positive slope during compaction until the point is reached when the flat portion of the punch contacts the already completely compacted portion of the bar which still remains under the punch surface. At this point, the pressing area increases sharply and thus the unit pressure actually decreases. A substantial increase in load is now required to raise the unit pressure to the level already achieved and thus to cause any further increase in green density. If the punch faces are faired properly, the surface of the pressed bar is also found to be very smooth and free of any marks at individual strokes.

The amount of advance per stroke may vary over a wide range. If it is too short, of course, the operation will be inefficient, and if too long, a smooth bar may not be obtained. In one experiment, which we conducted using a punch with a finishing area 1 in. long, it was found that varying the advance on an iron bar from $\frac{1}{3}$ in. per stroke to $\frac{3}{4}$ in. per stroke did not appreciably affect the green or sintered densities of the bar even with the same pressing load in both cases. One might, at first thought, expect an increased load to be required to achieve a given green density if the advance per stroke were increased. Consideration of the curve in Fig. 6 leads to an explanation of why this expected behavior is not noted in the experiment described above. The overlap per stroke which gives rise to the flat part of the pressing load versus punch travel curve provides a type of ballast to maintain uniform density once initial compact height is established.

Fig. 7. Iron-clad copper bar by continuous compaction.

The length of the "finish area" of the punch is a design factor which can be quite variable. It should be somewhat longer than the maximum desired advance per stroke and can be increased beyond this with no appreciable effect, except to require a greater load for the initial pressing stroke and also to make the punch more cumbersome.

Almost any kind of reciprocating press can be adapted to this process. A crank-operated punch press, for example, could probably be utilized. It is believed by the authors, however, that the most desirable type of mechanical press would be a cam-actuated type such as is commonly used for powder metallurgy work. Hydraulic presses of either conventional powder metallurgy design or of a wide variety of other reciprocating types can undoubtedly be adapted for use with this continuous compaction process.

The bar shown in Fig. 7 is iron-clad copper which illustrates one of the more interesting variations possible by continuous compaction. The uniformity of wall thickness should be noted. The bar in this illustration was made with one end entirely of iron and a copper center starting at some point in the mid-length and continuing through the bar.

Material	Compacting Load (tons)	Sintering Treatment	Pressed Size (inches)	Sintered Density (% Theoretical)	Results and Remarks
Copper	60	1 hr. @ 1000° C.	.25 x 2 x 28	90	Excellent bars and very easy to compact.
1-A (Repressed)	100	1 hr. @ 900° C.	.23 x 2 x 28	94	Slight surface marking from repressing.
2. Electrolytic Iron	60		.50 x 2 x 24		Good bars if powders are freshly annealed.
3. Easton RZ Iron	60	16 hrs. @ 1300° C.	.625 x 2 x 28	71	Excellent bars.
4. Tungsten + 1% paraffin	30	3.5 hrs. @ 1200° C. Plus	.50 x 2 x 24	61	Good appearance, but binder and low compacting pressure is necessary.
4-A Tungsten + 1% paraffin	40	3.5 hrs. @ 1600° C. 1 hr. @ 1200° C. Plus	.75 x .75 x 30		To be used as consumable electrode.
5. Molybdenum	15	3 hrs. @ 1650° C. 1 hr. @ 850° C. Plus	.50 x 2 x 24	82	Good bars. Low green density. Heavy shrinkage.
6. Columbium sponge	50	3 hrs. @ 1600° C.	.75 x .75 x 26		For use as consumable electrodes. Green bar is very flexible.
7. A-104 Iron sponge	50	15 hrs. @ 1150° C.	.75 x .75 x 26		For use as consumable electrodes. Green bar is very flexible.
8. Titanium sponge	50		.25 x 2 x 28		Compacts nicely. Used for "getter" sheets in sintering.
9. Nickel-Aluminum-Iron Alloy	95	4 hrs. @ 1125° C.	.50 x 2 x 28		Compacted well.
10. Iron-Cobalt-Vanadium Alloy	50	Pre-sinter ½ hr. @ 600° C.	.15 x 2 x 24	62	Excellent bars.
10-A (Repressed)	100	17 hrs. @ 1200° C.	.12 x 2 x 24	89	Slight surface marking. Subsequently processed to .004 tape.
11. Iron-clad Copper	60	1 hr. @ 850° C.	.30 x 2 x 12		Excellent bar. Special powder filling technique was necessary.

Fig. 8

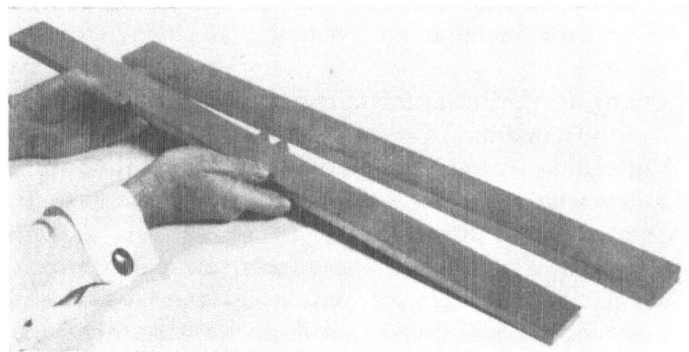

Fig. 9. Typical experimental bars.

Figure 8 lists some of the materials which we have processed by continuous compaction together with details of pressing, sintering, and further processing and some remarks on the finished bars. Copper powder pressed using a ram load of 60 tons and then sintered for 1 hr at 1000°C gave a very excellent bar of 90% density. Bars have been produced from various commercial types of iron powder including electrolytic, RZ, and Swedish sponge iron; all of which pressed satisfactorily.

In addition to straight metal powders, a variety of alloys were pressed from elemental powder mixes. These include iron—nickel magnetic alloys, iron—silicon alloys, KOVAR alloy (a Westinghouse proprietary iron—nickel—cobalt glass sealing alloy), and many others. The process has been used very successfully to press melting stock into electrodes for consumable arc melting. Titanium sponge, sponge iron melting stock, columbium roundels, and tungsten and molybdenum powders are among the materials prepared into electrodes for further melting.

Figure 9 illustrates the appearance of some experimental continuously compacted bars which were made in the 2-in. punch and die setup.

When any new process is developed, it is extremely interesting to consider the applications for which the process might be used immediately and to speculate on possible future applications. Some of the possible intermediate applications for this novel method of powder compaction are listed below:

1. Fabrication of large sizes of refractory metal sheet, strip or bar, for example, molybdenum, tungsten, tantalum, and so forth.
2. Continuous process for fabrication of tungsten from powder to filaments.
3. Continuous or semicontinuous fabrication of sheet or strip alloys where precise composition control is essential to obtain special properties.
4. Continuous fabrication into sheets, strip, or bar of alloys or metals which are too brittle as large ingots to be processed by conventional casting or working techniques. Beryllium is a fine example.
5. Low-cost fabrication of powder sponge or other compactable source material of highly reactive metals and alloys for further processing either by sintering and working or by vacuum arc melting the consolidated powder. This technique would enable high-purity, low-cost raw materials to be used and avoid problems now encountered, such as excessive segregation.
6. Production of continuous or semicontinuous ceramic bars.
7. Consolidation from powder of nuclear reactor core materials. These may be either metal, ceramic, or cermet.

A number of more speculative potential applications for our continuous compaction process can be visualized. The following are among the more interesting of such applications which come to mind at this time:

1. Single-step preparation of composite or clad bars, sheet, and strip of different metals. Examples of the above are thermal bimetals and clad reactor core elements manufactured in a single pressing operation. Bimetal bearing materials should also be considered in this classification.
2. Fabrication of bar, plate, sheet, and strip of common metals and alloys such as steel, copper, brass, etc., should become possible when low-cost sources of metal powders become available. Powder metallurgy fabrications of such common materials in competition with conventional processes does not seem too speculative when one considers the presently available low-cost sources of

copper and nickel powders. It is to be hoped that new iron powder production processes may result in substantial cost reductions for this raw material. Undoubtedly other applications will become apparent with time as a result of further investigation and utilization of this process.

In summary, a new technique for pressing powders into continuous bars is capable of producing pressed bars of many different metals and alloys, ceramics, and combinations thereof. The process described is unique in that relatively large cross-section bars of unlimited length can be readily produced.

Early experiments with this process indicate that it will provide an excellent method, economically and technically, of producing specialty metals and alloys in comparison to other fabricating methods. Relatively inexpensive, readily available equipment, e.g., reciprocating-type presses, can be easily modified to permit experimentation, development, and small-scale production of limited-length bars from powder, granular, or sponge raw materials. While manually operated equipment is adequate for experimental work, an automatic operation can be designed and built inexpensively for pilot plant work.

When lower-cost metal powder raw materials become available, the compaction process described in this paper will provide a basis for economical large-scale production of common metals and alloys in bar, sheet, and strip directly from powders.

Chapter 7.   Powder Rolling I

The Mechanism of the Compaction
of Metal Powders by Rolling

P. E. Evans

The Manchester College of Science and Technology
Department of Chemical Engineering
Manchester, England

A short historical review of powder rolling is followed by the description of some experiments on the rolling of copper powder. The factors which are of greatest importance, such as the rolling mill (roll diameter, size of roll gap, surface finish of the rolls, rate of powder feed to the rolls, and the roll speed), the powder (grain size and particle shape), and sintering (temperature, time, protective atmosphere) are taken into account of which the first two variables are mainly considered. On the grounds of the experimental results, the theoretical aspects of powder rolling are discussed.

Introduction

Formerly the size of products fabricated from metal powders was limited by press capacities, but if the powder is suitably fed between the rolls of a mill it may be compacted into a continuous strip. The realization of this technique during the last few years has greatly increased the potentialities of powder metallurgy.

The idea is not a new one. H. Bessemer, in his autobiography [1], remarks that in 1843 while rolling fine brass turnings
preparatory to making powder, the brass filaments tended to "issue from the rolls with a smooth continuous surface resembling
an ordinary sheet of solid brass." But he was interested in manufacturing powder not in compacting it, and it was not until 1904
that a deliberate attempt to compact metal powders by rolling
seems to have been made. In that year a patent was granted to
Siemens and Halske [2] in Berlin. It describes a method for compacting powders of the high melting point metals by feeding them
between two rolls whose axes are in the same horizontal plane. In
the years that followed, other patents for compacting by rolling
were granted but none had the simplicity of the Siemens–Halske
method. It does not seem to have been widely used, however, and
was next heard of in 1950 when G. Naeser and F. Zirm [3] published an account of their experiments on compaction by rolling.
In 1952, the author began a detailed experimental study of this
topic. A technique was devised whereby a standard two-high rolling mill could be used for compacting the powder [4] and the effects of the variable factors enumerated below were investigated
and correlated with the properties and microstructure of the strip
produced. The optimum properties shown by the strip compare
favorably with those of metal compacts made by the conventional
pressing techniques. A complete account of this work is contained
in a thesis submitted to the University of Cambridge [5] and papers
dealing with specific aspects of the problem have been published
elsewhere [6, 7].

Between 1952 and 1956, when the author's experimental work
was performed, a few papers devoted to compaction by rolling
were published, notably those by H. Franssen [8, 9], dealing with
various aspects of the rolling technique and production methods,
and those by S. Storcheim et al. [10, 11], describing the properties obtained under different experimental conditions. More recent papers are those by J. D. Shaw and W. V. Knopp [12] and by
D. K. Worn [13, 14]. Both deal with the engineering and econom-
ics of the process.

The purpose of the present paper is more general and fundamental in that it describes some of the phenomena that occur in
the continuous compaction process, and attempts to link them into

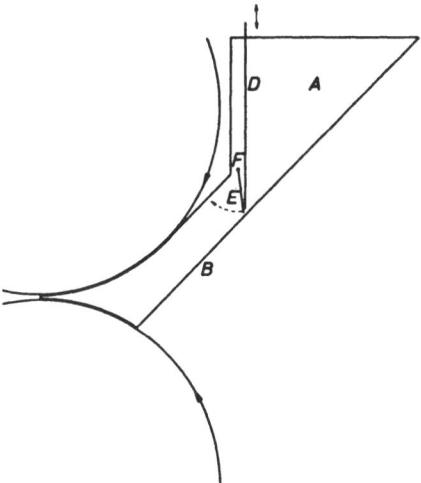

Fig. 1. Hopper feed to standard two-high mill.

a self-consistent explanation of the mechanism of compaction by rolling.

General Technique and Apparatus

The principle of the method which the author used for feeding powder to a standard rolling mill consists in allowing powder to slide down an inclined plane so that it is delivered to the lower roll sufficiently close to the plane of the roll axes for it to be carried forward into the roll gap by the motion of the rolls.

One form of the apparatus is shown in Fig. 1. The powder is contained in a hopper A, to which is joined a chute B, whose sides are mated at their lower ends to the roll profile. Accurately fitting gates D and E are fitted inside the hopper. The removal of D from its registering grooves allows the powder to flow through the variable aperture presented by E, which is pivoted at F. In this manner the rate at which powder is fed to the rolls may be varied. For powders which do not flow freely under gravity a vibrator may be attached at an appropriate part of the chute or hopper [4]. The width of the strip produced is determined by the width of the chute and is ultimately limited only by the barrel length of the rolls, assuming that there is enough power available to the mill.

A two-high mill with 8-in.-diameter rolls having a barrel length of $8\frac{1}{4}$ in. was used for the majority of the experiments. It was driven by a variable speed Schrage-type motor which had a maximum output of 12.5 hp. A "jeweller's mill" with 2-in.-diameter rolls was also available. The strip produced in this manner, which in this state is termed green strip, was batch-sintered in a Nichrome-wound electric resistance furnace through which dry, oxygen-free nitrogen was continuously passed.

Material

In experiments performed to test the range of application of the technique the metal powders successfully rolled into strip included iron, nickel, copper, titanium, tantalum, stainless steel powder, mixtures such as iron−copper, iron−chromium, copper −tin, copper−lead, silver−nickel, and metal and oxide mixtures such as copper−alumina. For the detailed investigation, however, copper was chosen because it is ductile, does not undergo allotropic modifications on heating, a wide range of powder particle sizes and a variety of particle shapes are readily available, and its sintering is a matter simple enough not to introduce doubts as to the efficiency of the compacting process.

The copper powders used may be broadly classified in four groups: electrolytic, water-atomized, air-atomized, and chemically precipitated powders. Each has a characteristic particle shape and structure (with the exception of water-atomized and air-atomized powders, whose structures are similar) which is determined by the method of manufacture. Electrolytic powder has a very irregular dendritic outline and shows twinned crystals; water-atomized powder has an irregular rounded outline and does not show twinning, air-atomized powder has almost perfectly spherical particles and shows no twinning. Unlike the preceding powders, each particle of the chemically precipitated powder appeared to be a single crystal, its shape intermediate between that of an electrolytic and a water-atomized particle.

Copper powders can be compacted by rolling even when they are heavily oxidized but the experiments were made with deoxidized powders except when the specific effect of surface oxide was being investigated. Deoxidation was effected by heating in hydrogen for 2 hr at 330°C.

Two densities may be quoted for a mass of uncompacted powder. They are: the apparent density, the ratio of the mass of powder to the volume it occupies measured immediately after the powder has been poured (under standard conditions) into the measuring vessel, and the tap density, the value of the same ratio after the vessel has been agitated until the powder occupies a minimum volume. For all powders used, except that chemically precipitated, the ratio tap density : apparent density was found to be 1.11 ± 0.08.

Variable Factors

The experimental program included, among other topics, the determination of the effects associated with the variable factors listed below. They fall logically into three main groups:

a. variables associated with rolling; that is, roll diameter, size of the roll gap, surface finish of the rolls, rate of powder feed to the rolls, and the roll speed;
b. variables associated with the powder, that is, particle shape, median size, and size distribution;
c. variables associated with sintering, that is, temperature, time, the nature of the protective atmosphere, and the method by which the material is heated.

Since the purpose of this paper is to elucidate the mechanism of compaction, the sintered strip will not be considered. Only the first two groups of variables are discussed below, and these only insofar as they yield information on what happens to the powder on its passage through the rolls. Similarly, only those properties of the green strip that give the most direct clues to the details of the compacting process fall within the scope of this paper.

Experimental Results and Discussion

Behavior of Powder in Roll Gap

Preliminary experiments showed that the powder, once it has been fed to the rolls, may be divided into two zones: one in which it is incoherent, and the other in which it exhibits an increasing degree of coherency up to the maximum dictated by the setting of the roll gap, Fig. 2. The extent of the two zones could be measured, to the nearest $\frac{1}{2}°$, by stopping the rolls before all the pow-

P. E. EVANS

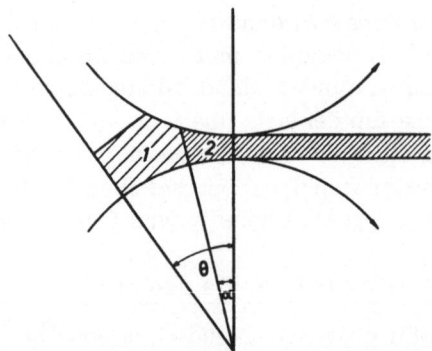

Fig. 2. Schematic representation of powder in the
roll gap. Zone 1, uncompacted powder; zone 2,
compacted powder.

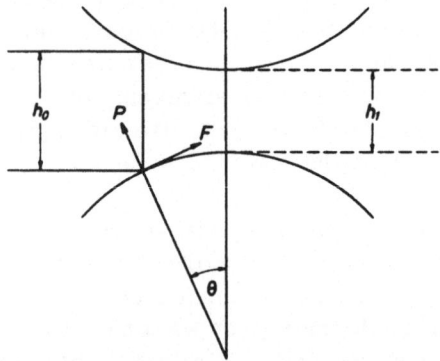

Fig. 3. Entry of a solid bar between the rolls.

der was compacted and carefully removing the feed apparatus. The
incoherent powder in the first zone was retained on the lower roll
for an angular distance from the roll gap determined by the angle
of rest of the powder. When this powder was removed by blowing
through a tube inserted from each side of the mill in turn, a coherent mass of powder forming the second zone remained, extend-
ing an angular distance α from the roll gap. The angle α may be
defined as the "gripping angle," a constant for a particular roll-
surface/powder combination.

It may be deduced from photomicrographs of the coherent powder in zone 2 that there is a gradual increase in the density of the powder as it passes through this zone, but the behavior of the powder in zone 1 is complicated, since it is undergoing some degree of agitation by the air expelled from between the particles ahead of it. Cine films of the rolling process, taken by the author show a continuous backward-tumbling motion of the powder in this region. Worn has shown that an increase in air pressure just above the powder mass may be detected with a manometer [13]. He also draws attention to the fact that the agitation effect is reduced if the air is replaced by hydrogen, which has a lower viscosity.

Rolling a Powder Compared with Rolling a Solid

Certain deductions may be made about the behavior of powder in the roll gap by comparing the rolling of powder with the rolling of a solid metal bar.

The entry of a solid bar between the rolls becomes impossible when the horizontal forces are equal to zero (Fig. 3); then,

$$P \sin \Theta = F \cos \Theta$$

i.e.,

$$\frac{F}{P} = \tan \Theta$$

but

$$\frac{F}{P} = \mu = \tan f$$

where f is the angle of friction between the bar and the rolls. Thus the bar cannot be drawn into the rolls when the contact angle Θ exceeds the friction angle f, Hence, the approximate value of f may be found in the following manner: for a bar of initial thickness h_0, held in gentle contact with the rolls, the roll gap is slowly increased from zero to the size at which the metal is just drawn between the rolls. Let the rolled thickness be h_1 (Fig. 3); then,

$$\frac{h_0 - h_1}{2} = R (1 - \cos \Theta) \qquad (1)$$

where R is the roll diameter. Maximum draft $(h_0 - h_1)$, occurs when $\Theta = f$.

For solid copper rolled between smooth steel rolls, the values obtained were: $f = 9°45'$ and $\mu = 0.17$.

For a certain electrolytic copper powder, the apparent density was 2.4 g/cc and the lowest density of coherent green strip was about 5.5 g/cc. Thus a compaction ratio h_0/h_1 of at least 5.5/2.4 is necessary in order to roll this powder into coherent strip. If it is assumed that the same value of f obtains for the powder as for the bar, that no slipping occurs between the particles, and that the pressure distribution and μ remain constant over the arc of contact, it is possible to calculate the theoretical maximum thickness of strip that could be rolled from this powder.

Substituting:

$$R = 101.5 \text{ mm}, \frac{h_0}{h_1} = \frac{5.5}{2.4} \text{ , and } f = 9° 45'$$

in (1),

$$\frac{5.5}{2.4} = \frac{2R(1-\cos f)}{h_1} + 1, \text{ and } h_1 = 2.2 \text{ mm.}$$

It was found by experiment, however, that this powder gave a strip 0.8 mm thick at a density of 5.5 g/cc. Substituting this value of h_1 gives $f = 5°48' \approx 6°$, which is the same figure as was obtained, by direct measurement, for the gripping angle for this powder.

The discrepancy between the calculated value of f and that obtained for the solid bar shows that the assumptions made above are not valid. If the variations in pressure and μ over the arc of contact are ignored, being of secondary importance, it means that during rolling, slip between the particles occurs, at least up to that point on the roll surface that marks the beginning of the coherent zone. Even within the coherent zone it is not necessary to postulate that all relative particle movement has ceased, the coherency merely implies that more metal-to-metal bonds are formed in this region than are broken. This is considered in a later section.

The foregoing elementary analysis makes it desirable to examine the concept of "compaction ratio" as it applies to compac-

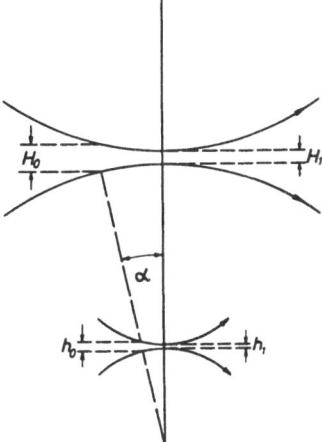

Fig. 4. Effect of roll diameter on strip thickness when compaction ratio $h_0/h_1 = H_0/H$, and gripping angle α constant.

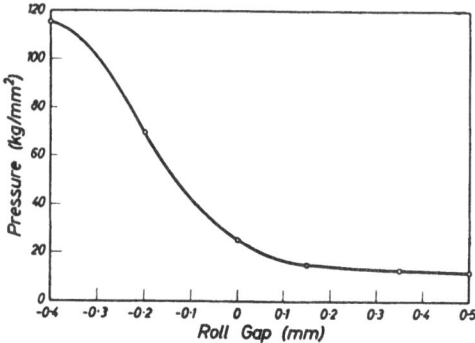

Fig. 5. Variation of pressure on powder with varying roll gap (negative values of roll gap denote mutual roll pressure before rolling began).

tion by rolling. In standard powder metallurgy practice, the term refers to the ratio of the depth of powder in the die before compaction to the depth after compaction. Since the mass of powder is constant, this is equal to the ratio of the final density to the initial density. A similar definition for powders compacted by rolling would be the ratio of the density of the green strip to the minimum density of the powder in the coherent zone.

Table I

Surface Finish	Strip Thickness
Highly polished	0.35 mms.
Smooth matte	0.55 mms.
Shot blasted	0.70 mms.

Table II

Powder	Gripping Angle
Electrolytic	$6° \pm \frac{1}{2}°$
Water-atomized	$3\frac{1}{2}° \pm \frac{1}{2}°$
Air-atomized	$1\frac{1}{2}° \pm \frac{1}{2}°$

Table III*

Water-atomized Powders			
Type	Percentage of particles passing through sieve openings of sizes shown (microns)		
A	91%	240—175	9%	175
B	95%	150—140	5%	140
C	80%	100—85	20%	85
D	78%	75—60	22%	60
E	100%	44	—	—

*See also Table III, p. 133.

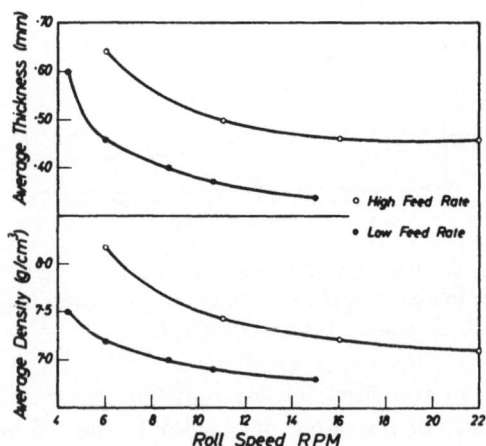

Fig. 6. Effect of roll speed on thickness and density of green strip.

Rolling Variables

With all other conditions constant, the effect of using rolls of different diameter may be deduced from Fig. 4. For strip rolled to a given density, and for a constant gripping angle, the compaction ratios are equal, that is, $h_0/h_1 = H_0/H_1$, and since $H_0 > h_0$, $H_1 > h_1$. Thus, the thickness of strip of a given density should increase as the roll diameter is increased. This has been substantiated by G. Naeser and F. Zirm [3], and by the author.

It may also be seen from Fig. 4 that as the roll gap is increased, the compaction ratio decreases. Eventually a point is reached where the powder is no longer coherent. The variation of the average rolling pressure with varying roll gap is shown in Fig. 5 for an electrolytic powder rolled at 24 rpm.

The thickness of strip of a given density increases as the surface roughness of the rolls is increased. Three pairs of rolls were available for the 8-in. mill. Each pair was given a different surface finish: shot blasted, smooth matte, and highly polished. At a given density the thickness of strip was as shown in Table I.

The increased friction between rolls and powder has led to an increase in the gripping angle, so that for a constant compaction ratio, that is, a given rolled density, the thickness of strip is increased.

In compaction by rolling there are three rate-controlling elements: (1) the hopper aperture – which determines the rate at which a given powder reaches the rolls; (2) the roll speed and roll gap – which determine the rate at which a given powder is compacted into strip; and, (3) the flow properties of the powder which for a given roll speed and gap determine the rate at which the powder passes from the incoherent to the coherent zone. It is convenient to define two feed rates: (a) the "external" feed, is determined by the aperture of the hopper and the flow properties of the powder; (b) the "internal" feed, is determined by roll speed and gap, and the flow properties of the powder. The superincumbent weight of powder in zone 1 depends, in an obvious manner, on the relation between these three elements.

When, for a given external feed rate, the roll speed is increased, the thickness and density of the strip produced at a fixed setting of the roll gap decreases in the manner shown in Fig. 6. When the external feed rate was increased fourfold, a similar decrease occurred but the thickness and density of the strip were greater than the values obtained for the lower feed rate at any given roll speed. The powder supply became inadequate at speeds above about 15 rpm for the lower feed rate and above 25 rpm for the higher feed rate. The powder then passed through the rolls without being compacted into strip. When this powder was compacted in a die, the density obtained at a given load was about 12.5% higher for a time-under-maximum-load of 30 sec than for a time of about 1 sec. For a change in roll speed from 15 rpm to 6 rpm, there was a 7% increase in density at the low feed rate and a 14% increase at the high feed rate. Hence, the decrease in density of the strip as the roll speed is increased may be ascribed in part to the shorter time for which each elemental length of strip is subjected to pressure and in part to a greater agitation effect on the powder in zone 1 due to a higher velocity of escape of entrapped air.

The decrease in thickness with increasing roll speed probably occurs because the internal feed is decreasing. The increase in thickness and density with increased external feed is due, probably, to an increase in the density of the incoherent powder (due to less agitation) as it passes through zone 1.

It may be concluded that for an internal powder feed that just keeps pace with the rolling speed, the powder will pass through zone 1 at a low density, approximately equal, say, to its apparent density. On the other hand, for a high internal feed, the superincumbent weight of powder would tend to reduce the agitation produced by escaping air and a more dense strip should be obtained.

Powder Variables

The attributes of a powder which directly influence its behavior in rolling would seem to be the particle shape and the average particle size. On these depend the thickness and density of strip produced under fixed rolling conditions. The relationship may best be interpreted in terms of the effect of these variables on the gripping angle and the powder flow, which regulates the powder feed.

Samples of powders of different particle shape but the same size distribution showed the gripping angles listed in Table II.

The variation in gripping angle with the type of powder used may be mainly attributed to the friction between the powder and the rolls, since (water-atomized) powders of the same general shape but of different average size (see Table III) showed the same gripping angle, within the limits of error ($\pm\frac{1}{2}°$) to which it could be determined.

It was initially thought that strip thickness at a given roll setting depended only on the gripping angle but for the powders listed in Table III it was found that the thickness of strip produced under fixed rolling conditions decreased with decreasing average particle size.

As the average particle size decreases, the surface area per unit mass of powder increases. It may be concluded that strip thickness is controlled both by gripping angle and by the particle size. This experimental evidence and the difference between the gripping angle for solid copper and that for the powder, plus the fact that air-atomized powder could not be compacted into strip by rolling and the observation that a very fine powder with an average particle size of about 10 μ could only be compacted at very low rolling speeds, prompted a study of the relationship between the flow properties of a powder and its behavior in rolling.

A measure was sought of a characteristic of a mass of particles in relative motion that was thought to be time dependent and dependent on interparticle contact area, that is, analogous to viscosity rather than to friction. This was achieved by measuring powder flow rates from a hopper of similar design to that previously described. The experiments showed that the flow of a mass M g of copper powder through an aperture of area A mm^2 measured perpendicular to a chute inclined at an angle Θ to the horizontal took place in a time t sec, such that:

$$t = k \cdot \frac{M}{A^n \sin \Theta}$$

where k and n are constants for a given powder. The mean value of n for powders of different shape and mean size was 1.55 ± 0.03. The term k is the viscosity analog sought in these experiments. It

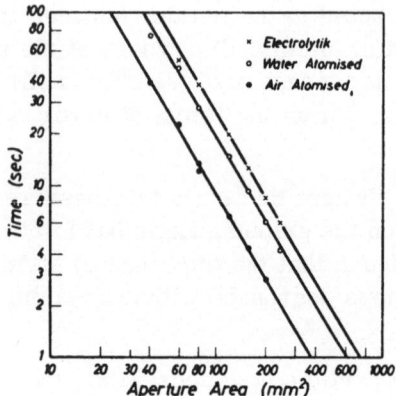

Fig. 7. Relationship between log flow-time
and log aperture-area when M = 800 g and
θ = 60°.

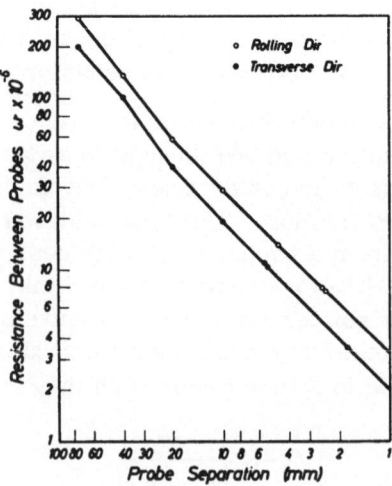

Fig. 8. Resistance in the rolling and transverse
directions for different probe separations (deoxidized powder).

is referred to below as the "powder viscosity factor." The analogy seems justifiable, since the communication or impedance of
motion between particles is more akin to viscosity than to friction.
The difference between this flow formula and that arrived at by
C. J. Leadbeater et al. [15] is due only to the different geometry
of the apparatus.

Further experiments showed that k increased with increasing
irregularity of the particle surface of the powder. It was, for example, found to be in the ratio $1:2:3$ approximately, for air-
atomized, water-atomized, and electrolytic powders, respectively,
which had a certain common size distribution. A typical set of
curves is reproduced in Fig. 7. For particles of a given shape, k
increases as the average particle size decreases, the increase being more pronounced for average particle sizes less than 100μ.
The correlation between the powder viscosity factors of a series
of powders of the same general shape (i.e., constant gripping
angle) and the thickness of strip produced from them was fair, but
a heavily oxidized powder showed an anomalous (increased) thickness. This effect may be due to increased roll particle friction
similar to that obtained with solid metals though no corresponding
increase in gripping angle could be detected. It was probably below the limits of error.

Thus, the viscosity factor of a particular powder provides
some indication of how it is likely to behave in compaction by rolling. It is inversely proportional to the flow rate of the powder and
and the flow rate : roll speed ratio determines the density and
thickness of the strip, when other factors are constant, as described above.

The term "friction" is reserved in the present context for the
tangential force opposing the relative motion of contiguous bodies,
i.e., it is applied to roll/particle and to particle/particle combinations but not to the mass of powder as a whole.

Although they were suitable for compacting in a die, the very
fine, chemically precipitated powders (slow flowing or high k)
could only be compacted in rolling at very low rolling speeds,
whereas the faster-flowing electrolytic and water-atomized powders with lower k values could be compacted over a range of
speeds. When, however, the value of k fell below about 24, as de-

termined in the manner described above, the powder could no
longer be compacted by rolling. The experiments indicated that
such a low value of k was only found for spherical particles. It
must be concluded that the degree of compaction experienced under
these rolling conditions by a mass of spherical particles does not
allow the formation of sufficient metal-to-metal bonds per unit
volume of strip for the strip to be capable of supporting its own
weight. Hence, for a particular powder to be a feasible material
for continuous compaction, it must have good flow properties and
be composed of particles whose shape-size distribution and ductility will provide enough intimate points of contact to ensure a coherent green strip.

Resistance Measurements

In order to obtain more information about the behavior of powder particles during rolling the electrical resistance along different directions in the plane of the green strip was measured with a
Kelvin Bridge having a range from 0.1 $\mu\Omega$ to 1.1 Ω. Copper has
an isotropic electrical resistance, so that any anisotropy in the
resistance of compacted powder must be caused either by local
variations in density or by differences in the degree of bonding in
different directions in the sheet. The resistance of unsintered
strip is very sensitive to variations in density, particularly at low
densities, and there is always the possibility that this effect may
mask the other.

After a number of preliminary experiments, the two effects
were separated. Specimens 10 cm^2 were used and the resistance
between probes of variable separation was measured along two in-
tersecting lines in the rolling and transverse directions in turn.
As may be seen from Fig. 8, the relation between probe repara-
tion and resistance is linear, on a log-log scale, for separations
less than about 2.0 cm. By extrapolation, Fig. 8 shows the difference in the resistance in the two directions for a region about 1
mm^2 at the point of intersection of the two lines. The results of
measurements on numerous specimens showed that the resistance
in the transverse direction was lower than that in the rolling direction. Some of the same batch of deoxidized electrolytic powder
as that used for the specimen of Fig. 8 was then heavily oxidized
by heating in air for several hours at 100°C. After a specimen
had been prepared and measurements made, under exactly the

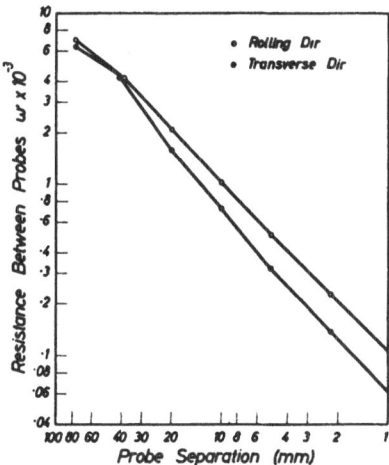

Fig. 9. As for Fig. 8, but strip rolled from
heavily oxidized powder.

same conditions as before, the resistance in the rolling direction
was found to be 88% higher than in the transverse direction, Fig. 9.
The corresponding figure for the deoxidized powder was 60%.

It was suggested above that slip occurs during rolling at least
up to the point at which the powder is gripped by the rolls. The resistance measurements show that relative motion of the particle
surfaces, if not of entire particles, occurs even within the second
zone where the particles are in sufficiently close contact for abra-
sion to take place. Relative surface motion in the rolling direction
would help to remove surface films from those parts of the particles parallel to the rolling direction more than those in the transverse direction, thus giving a greater number of metal-to-metal
bonds per unit area across planes parallel to the rolling direction.
If the proposed mechanism is correct, the effect should be more
pronounced when the particles are surrounded by a heavy oxide
layer. This is verified by the experimental evidence.

Conclusions: The Mechanism of Compaction

Any elaboration of a mechanism of compaction must explain
the following phenomena: (a) a given powder is not compacted if
the roll gap is too large; (b) increasing the powder feed increases

the thickness and density of strip produced at a given roll gap; (c) increasing the roll speed decreases the thickness and density of strip produced at a given roll gap; (d) the green strip shows better bonding across vertical planes parallel to the rolling direction than across vertical planes in the transverse direction; (e) the gripping angle varies for particles of different general shape; (f) the gripping angle is approximately constant for particles of a given general shape and varying particle size, but the thickness of the strip produced decreases with particle size.

From the experimental evidence summarized above, the following explanations may be advanced:

a. The powder must undergo a minimum degree of compaction if a coherent strip is to be produced. For a given powder the compaction ratio is determined by the gripping angle and the roll gap and, since the gripping angle is fixed, increasing the roll gap decreases the compaction ratio until eventually a point is reached where the powder is not compacted.

b. Increasing the powder feed increases the packing density of the powder in the roll gap by reducing the agitation of the powder so that for a given compaction ratio the density of the strip is increased. As the density is increased, the force separating the rolls must increase, resulting in roll flattening and a thicker strip.

c. Increasing the roll speed reduces the time during which the powder is subjected to a compressive load, thus decreasing the density of the strip. Furthermore, as the roll speed increases, the velocity of escape of entrapped air must increase and the superincumbent weight of powder must decrease, so that the powder density is decreased. This, too, reduces the density of the strip.

d. It is clear from the results of the electric resistance measurements that the particles or particle surfaces undergo relative movement in the rolling direction. Relative movement of the particle surfaces alone could only occur when the particles were already intimately packed and undergoing deformation in the rolling direction. On the other hand, relative motion of discrete particles might take the form of a rotary motion induced by the couple formed by the friction between the rolls and the adjacent layer of particles and the constriction of the roll gap. Relative motion would then be greater between horizontally adjacent particles than between vertically adjacent particles, which would tend to "gear" together, and there would then be a more vigorous abrasive effect in planes parallel to the rolling direction than in the other two principal planes. This would tend to remove surface films and so permit improved bonding to occur across the one plane.

e. The rotary motion would be communicated through the powder and slipping between the particles would give rise to a powder viscosity effect whereby the force along surfaces concentric with the roll surfaces would decrease as the distance from the roll surfaces increased. For increasingly irregular powder surfaces, the effect would persist for an increased distance from each roll surface and the gripping angle would be determined by the point in the roll gap where the rotary motion imparted to the powder by one roll surface met that imparted to the powder by the other roll surface. Thus, the gripping angle would increase with increasingly irregular particle surfaces.

f. To a first approximation the gripping angle is determined by the roll-powder friction, which in turn has been shown to be dependent on powder-particle shape. The only variable associated with particle size is the powder viscosity factor and it would appear that, in the absence of variations in particle surface conditions capable of affecting the roll-powder friction, the powder viscosity factor determines the strip thickness – mainly by its control of the internal powder feed rate.

Acknowledgment

The experimental work was carried out at the Goldsmith's Laboratory of the Department of Metallurgy in the University of Cambridge. The author is grateful to Mr. G. C. Smith, M. A., and and to Professor G. Wesley Austin, O. B. E., for their continuous interest and encouragement and for many helpful discussions during the course of the work.

118   P. E. EVANS

References

1. Sir Henry Bessemer, F. R. S., An Autobiography, Engineering, London (1905), p. 68.
2. D. R. P. 154,998.
3. G. Naeser and F. Zirm, Stahl Eisen 70:995 (1950).
4. British Provisional Pat. No. 34541/53.
5. P. E. Evans, Ph. D. dissertation, The Continuous Compaction of Metal Powders, Cambridge University (1956).
6. P. E. Evans and G. C. Smith, Powder Met. No. 3:1 (1959).
7. P. E. Evans and G. C. Smith, Powder Met. No. 3:26 (1959).
8. H. Franssen, Metall 8:365 (1954).
9. H. Franssen, Metal Ind. 86:227 (1955).
10. S. Storcheim, J. Nylin, and B. Sprissler, The Sylvania Technologist 3(2):42 (1955).
11. S. Storcheim, Metal Progr. 70(3):120 (1956).
12. J. D. Shaw and W. V. Knopp, Proceedings of the Thirteenth Annual Meeting, Metal Powder Association, Chicago, Vol. 1 (1957), p. 33.
13. D. K. Worn, Powder Met. No. 1/2:85 (1958).
14. D. K. Worn and R. P. Perks, Powder Met., No. 3:45 (1959).
15. C. J. Leadbeater, L. Northcott, and F. Hargreaves, Selected Government Research Reports, Vol. 9, Powder Metallurgy, H. M. S. O., London (1951), p. 33.

Chapter 8. Powder Rolling II

The Compaction of Metal Powders by Rolling

P. E. Evans
Department of Chemical Engineering
Fuel Technology, and Metallurgy
Manchester College of Science and Technology
Manchester, England
and
G. C. Smith
Department of Metallurgy
University of Cambridge
Cambridge, England

The Properties of Strip Rolled From Copper Powders

The investigation described is an extension of earlier work ["Symposium on Powder Metallurgy 1954," (1956), p. 131; London, Iron and Steel Institute, and Sheet Metal Ind. 32 :589 (1955)] which described the effect of rolling pressure and sintering conditions on the mechanical properties of strip rolled from copper powders. The directional variation of UTS and of elongation of sintered strip are shown to be the same as those of solid copper with a similar microstructure, at least for material with up to 16% porosity. The shape of powder particles and the particle-size distribution have a marked effect on the strength of sintered

119

strip by virtue of their effect on the shape and size of the pores in the sintered material. Measurements of electrical resistance reveal a linear relationship between conductivity and porosity over wide ranges of porosity in both "green" and sintered strip. The conductivity increases rapidly during the first few minutes of sintering at 1000°C. Measurements of the resistance in the rolling direction and in the transverse direction, which are independent of local variations in density, have been made on green sheet; the resistance in the rolling direction is the higher. A correlation between this result and the mode of particle deformation is proposed, and is elaborated in the second part of this paper (p. 148).

I.   Introduction

The details of the method employed by the authors in rolling strip directly from metal powders by feeding them from a hopper via a chute to a standard two-high rolling mill (8-in.-diameter rolls), and a survey of the effects of some rolling and sintering variables on the properties and structure of such strip, have been given elsewhere [1]. The historical background to powder rolling and the results obtained by other workers — notably Naeser and Zirm in Germany — in comparison with these findings were discussed in a second paper [2].

The essential difference between compacting metal powder in a die and compacting by rolling is the possibility of a large amount of anisotropic deformation during rolling, which cannot occur during compaction in a die. For this reason the experiments described earlier [1] were designed to determine whether there was any marked difference in properties in the two cases. It was found that the strength and ductility of strip prepared by rolling and sintering powder were at least as high as those obtained by pressing and sintering under similar conditions. Measurements have now been made of the variation in tensile strength with direction in the sheet, and the effects of powder variables on the strength of sheet produced under fixed rolling and sintering conditions have been determined. Finally, the electrical resistance of the sheet in different conditions and in different directions has been measured.

II. Variation of Tensile Strength with Direction

1. Earlier Work

Directional properties in cast and worked metals may arise from three principal causes: crystallographic preferred orientation, the elongated shapes of the grains themselves, and the preferred orientation of inclusions – a term which here embraces elongated cavities, blow-holes, and segregates as well as nonmetallic inclusions. In the experiments described below, all the tensile specimens were prepared from fully sintered material (2 hr at 1000°C) showing substantially rounded pores. They were annealed after preparation, so that any directional properties should be attributable to crystallographic preferred orientation rather than to the other sources mentioned above.

The effect of rolling and annealing upon the crystallography, metallography, and physical properties of dense copper strip has been investigated by Baldwin [3], among many others. Earlier, Brick and Williamson [4] had studied the cold-rolled structure of copper strip, but a more recent, quantitative analysis of rolling texture has been made by Hu, Sperry, and Beck [5]. The latter found their results to be in close agreement with an ideal orientation of the $(123):[1\bar{2}1]$ type.

So many states of the material can be induced by different combinations of cold-rolling and annealing processes that Baldwin's work, which relates directional properties to crystallographic structure, would seem to offer a more fruitful approach than an attempt to correlate directional properties directly with cold-rolling and annealing conditions. A random crystallographic orientation of grains gives properties that are very nearly isotropic, whereas varying degrees of cubic alignment [6] of grains – $(100):[100]$ in the rolling plane and rolling direction – above about 60%, give directional properties characterized by an elongation/direction curve that rises to a maximum at about 45° to the rolling direction.

2. Experimental

The copper powder (Table I) was rolled directly to sheet in the manner previously described [1,2]; sintering was carried out in a nitrogen atmosphere and tensile tests were made with a

Table I. Powder Used for Determining
Directional Properties

Sieve No.	Size, μ	Wt.-%
+100 mesh	>147	0·2
−100 +200	147–74	21·6
−200 +325	74–44	36·7
−325	44	41·5
Sub-sieve range:		
	−25	14·6
	−15	4·0

Specific surface, cm.²/g.: 520
Apparent density, g./c.c.: 2·85

Chevenard microtensile machine on specimens, with a 5-mm gauge-length, that had previously been given a standard anneal in vacuum for 1 hr at 450°C.

The experiments were designed to determine:

(a) The variation of tensile properties with direction for sintered strip annealed after the preparation of the test pieces.

(b) The variation of tensile properties with direction for sintered strip subsequently reduced 75% by cold rolling, and then annealed after preparation of the test-pieces.

(c) Whether preferred orientation exists in green strip, and the nature of the preferred orientation of the cold-rolled but unannealed material prepared as in (b).

The details of the experimental methods adopted for each of these three groups are as follows:

(i) The powder was rolled, with an initial roll gap of 0 mm and a roll speed of 6 rpm, into strip 15 cm wide. The center material − a strip 2.5 cm wide − was removed and sintered for 2 hr at 1005° ± 5°C in nitrogen. Square specimens were cut from the full width of this strip and their densities measured. From those squares with densities lying within the range 7.3− 7.6 g/cc, blanks for tensile specimens were cut in directions making angles of 0°,

$22\frac{1}{2}°$, 45°, $67\frac{1}{2}°$, and 90° with the rolling direction. The densities of the blanks were again measured, and the specimens grouped into three density ranges: 7.4 ± 0.05, 7.5 ± 0.05, and 7.6 ± 0.05 g/cc, so that in each range the density could be considered constant within the limits of experimental error. Tensile specimens were prepared from these blanks and given the standard annealing treatment of 450°C for 1 hr in vacuum. For each direction, quoted with respect to the rolling direction, at least three specimens were tested.

(ii) Strip prepared as described in (i) was reduced 75% by cold rolling. This appeared to be near the upper limit of cold reduction, since edge cracking occurred to a depth of about 3 mm, but only the center part of the strip was needed. The density of fused copper was obtained by this reduction. Blanks for tensile specimens were cut from the strip and prepared as described above.

(iii) Qualitative pole figures were determined from specimens whose thickness had been reduced to 0.004 cm by dissolution in 50% nitric acid.

3. Results and Discussion

The variations of ultimate tensile stress and elongation with direction in the sheet are plotted for once-rolled and sintered material in Fig. 1. The change in UTS with direction is only about 4% which, under the conditions of testing, is approximately equal to the maximum error involved in the use of the Chevenard machine [7], so that very little significance attaches to it. The change in percentage elongation would seem to be more variable, judging by the results for the material of density 7.5 g/cc. However, the elongations for the directions 0°, 45°, and 90° exhibit little variation in all three groups.

The microstructure of this material shows that the percentage of cubically aligned grains is very small. Thus, material having an average density of 7.5 g/cc has directional properties very similar to those of fully dense material having the same randomness of grain orientation.

For material that has been cold-rolled 75% and annealed at 450°C for 1 hr after the initial rolling and sintering treatment, the

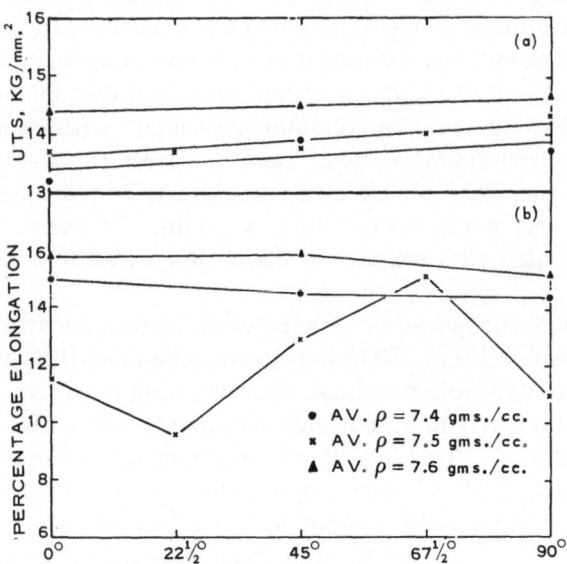

Fig. 1. Variation of (a) UTS and (b) elongation with angle
to rolling direction in rolled and sintered copper strip.

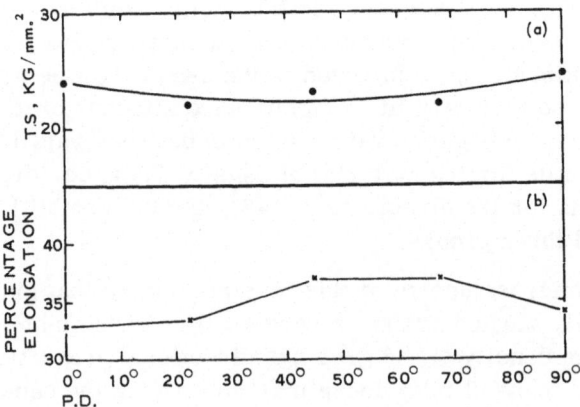

Fig. 2. Variation of (a) UTS and (b) elongation with angle
to rolling direction in copper sheet cold-rolled 75% and annealed.

variations in UTS and elongation are plotted in Fig. 2. For the
conditions under which the experiment was carried out, the variations in percentage elongation do not differ radically from those
found by Baldwin [3] for copper strip having a low degree of cubic
alignment. The small amount of cubic alignment found in the present strip does not preclude the possibility of some degree of preferred orientation, which is indicated by the significant variation
in elongation.

The microstructures of strip (reduced 75% in cold rolling) before annealing did not show any pronounced cubic structure. That
such a structure is not in fact present is borne out by the pole figures, compiled from x-ray diffraction patterns of this material.
The typical structure of this material, after it had been annealed
at 450°C for 1 hr in vacuum, again shows no pronounced cubic
alignment. A few traces of twin planes making angles of 45° with
the rolling direction could be seen, but not the continuous matrix
of cubically aligned grains which Baldwin has shown to be necessary to produce the characteristic pronouned maximum in the
elongation at an angle of 45° to the rolling direction. This further
confirms that, as with cast, rolled copper, the directional properties of strip rolled from copper powder are governed by the orientation pattern of the grains, the voids being isotropic in their
effect on the tensile properties.

X-ray evidence indicates that no marked orientation exists in
green strip rolled from the given powder with an initial roll gap
of 0 mm; that this state, with some grain growth, persists after
sintering; but that after a reduction of 75% in cold rolling, a high
degree of preferred orientation is present in the center of the
strip. The pole figures show that this preferred orientation is
qualitatively the same as that found at the center of heavily reduced cold-rolled solid copper. After the standard anneal, to
which all tensile specimens were subjected, there was no marked
preferred orientation. The absence of a high degree of preferred
orientation in the sintered strip does not necessarily mean that
earing will not occur in deep drawing. Baldwin has shown [3] that
the preferred orientation corresponding to only 30% of cubically
aligned grains can still produce an ear height of 5%. Thus, the
suitability of this material for drawing will depend on its microstructure, which must be controlled by the usual methods.

Table II. Powders Used for Determination of Effect
of Size Distribution on Strip Properties

Type of Powder	Designation	Percentage of Particles Passing through Sieve Openings of Sizes Shown						Apparent Density, g./c.c.
		%	μ	%	μ	%	μ	
Electrolytic	E1	80	384–147	15	147–44	5	44	3·50
	E2	10	208–147	80	,,	10	,,	3·10
	E3	5	384–147	15	,,	80	,,	2·73
Water-atomized	W1	69	351–147	27	147–44	4	44	3·55
	W2	9	208–147	64	,,	7	,,	3·50
	W3	10	351–147	15	,,	75	,,	4·16

The pole figures of the forms {111} and {100} obtained from
strip cold-rolled 75% are in close qualitative agreement with
those found by Hu, Sperry, and Beck [5] for the inside texture of
tough-pitch copper reduced 96% in cold rolling.

It may be concluded that strip rolled from copper powder
under the conditions described has tensile properties in different
directions that are similar to those of solid copper with the same
microstructure. This is independent of the density of the strip for
densities as low as 7.5 g/cc.

III. Powder Variables

1. Earlier Work

Preliminary experiments have shown that, of the copper powders most readily available – air-atomized, water-atomized, and
electrolytic – only the latter two can be easily compacted to form
strip. There were no grounds for deducing whether the irregular,
electrolytic powder particles or the water-atomized powder particles, which have a more regular outline, would yield the better
mechanical properties when rolled and sintered under fixed conditions. Furthermore, the effect of size distribution in the powder is a variable factor that may affect the properties of the strip.
Experiments were made with electrolytic and water-atomized powders in order to establish the effect, if any, of the two characteristic particle shapes. A normal size distribution and two skew
distributions with modal particle sizes of medium, large, and
small particles, respectively, were chosen for each powder type.

The results of these experiments led to the investigation of the effect of pore size and grain size on the strength of sintered strip.

2.   Experimental Results and Discussion

(a)   The Effect of Particle Shape, Size, and Size Distribution

Details of the electrolytic and water-atomized powders used in these experiments are given in Table II.   The size distributions show a fair degree of correspondence for the two different particle shapes.   All the powders were rolled with an initial roll gap of 0 mm at a speed of 4-6 rpm and sintered for 2 hr at $1010° \pm 5°$ C in nitrogen.   The density and tensile properties were determined in the usual manner [1].

Examination of etched sections of the sintered strip showed the following general characteristics: (a) the average grain size decreased with decreasing initial particle size; (b) the total porosity increased as the average initial particle size decreased; (c) the large pores were associated with a large initial particle size, an intermediate pore size with the smallest initial particle size, and the smallest pores with the strip rolled from the powders having a normal size distribution; (d) the pores in the strip rolled from electrolytic powder were rather more angular than those in the strip from water-atomized powder.

The UTS and the percentage elongation for sintered strip are plotted against the sintered density in Fig. 3.   In Fig. 4 these properties are plotted against the approximate average particle size of the powders.

For the range of densities obtained in these experiments, the UTS and percentage elongation are greater for the strip rolled from water-atomized powders than for the strip rolled from electrolytic powders at any given density.   This result could not have been predicted and justifies further consideration.

Scheuer [8], investigating the effect of blow-holes of various shapes on the mechanical properties of cast aluminum alloys, obtained tensile strengths of 22.5, 21.2, and 16.5 kg/mm^2 for specimens with round blow-holes, no blow-holes, and "dendritic" blow-holes, respectively.   Under the present rolling and sintering conditions there is a general tendency, remarked above, for the sintered strip to show a more angular porosity when prepared

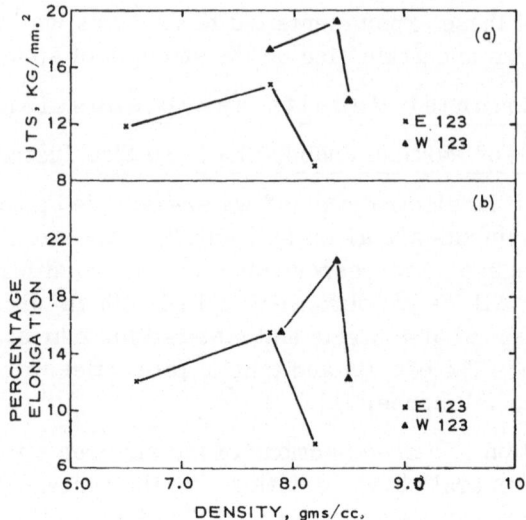

Fig. 3. (a) UTS and (b) elongation of sintered strip plotted
against density.

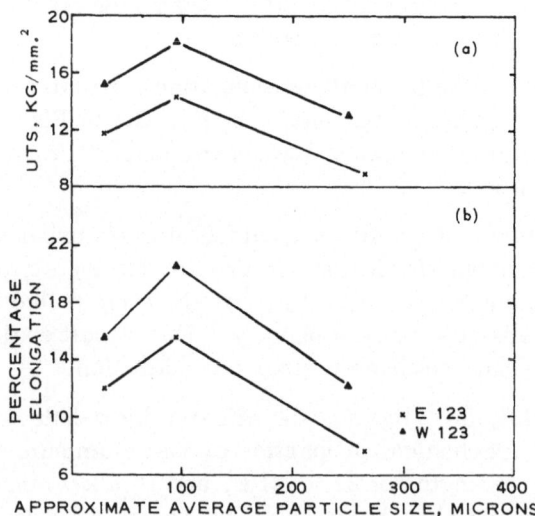

Fig. 4. (a) UTS and (b) elongation of sintered strip plotted
against the approximate average particle size of the powder.

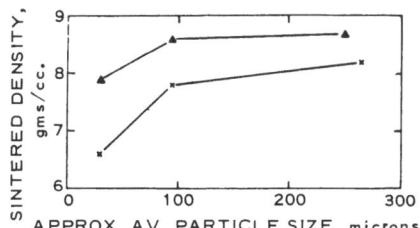

Fig. 5. Density of sintered strip plotted against
the approximate average particle size of the
powder. ▲ – W 1 2 3; × – E 1 2 3.

from an electrolytic powder than when a water-atomized powder
is used. This might account for the greater strength at a given
density of the strip rolled from the latter.

However, another characteristic of the curves of Figs. 3 and
4 demands further investigation.

Both the UTS and the elongation show maxima when plotted
against either density or approximate average particle size. This
evidence suggests that two opposing factors are at work. One is
the maximum pressure exerted on the powder during rolling. For
a given setting of the roll gap the rolling pressure (and hence the
density) increases with particle size up to a certain maximum particle size and thereafter remains fairly constant. On sintering,
this general trend of the density/particle-size relationship persists (Fig. 5), thus accounting for the increase in strength and
elongation as the average particle size increases. If this were the
only factor involved, the tensile properties should also increase
with particle size and then remain constant. However, the curves
of Fig. 4 show that, beyond a certain particle size, the tensile
properties decrease again. This decrease is independent of the
total porosity, as shown by the curves of Fig. 3. It is suggested,
therefore, that the decrease in tensile properties is caused by the
increased pore size and grain size associated with the increased
average size of the original particles. It has already been observed that the average grain size increases with increasing initial particle size; there remains to be determined the effect of initial particle size on the pore size of strip rolled and sintered
under fixed conditions and the effect of the pore size of such strip
on its tensile properties. This is considered at length below.

(b) The Effect of Pore Size and Grain Size on the Strength of Sintered Strip

It was suggested above that a greater pore size decreases the
tensile strength of strip of a given density. Whereas it seems to
have been generally realized that large pores reduce the strength
of sintered compacts, no quantitative evidence has been advanced
in support of this view. Goetzel states simply that the detrimental effect of maximum pore size is usually more important than
total porosity [9], and that the pore size can be controlled by the
type of powder used [10]. For cast bronze, Pell-Walpole [11] obtained a higher UTS for a casting with a uniform distribution of
pores than for the same percentage porosity occurring as a concentration of residual shrinkage pores along the axis of the test
specimen. In interpreting this relationship, however, he does
not seem to have taken the pore shape into account. The residual
shrinkage porosity was probably the same as that described by
Scheuer [8] as "dendritic," while the evenly distributed pores were
attributed to gas being absorbed during casting and were probably
nearly spherical.

Some attention has been paid to the effect of grain size on
tensile properties [12] and Carreker and Hibbard [13] have shown
that whereas, for solid copper, the UTS is independent of grain
size, the yield strength decreases as the grain size is increased.
Subsequently, different rates of strain-hardening allow the specimens to approach the same flow stress at high strains. If the
yield stress is independent of the presence of pores, i. e., if the
yield stress of coarse-grained sintered material is lower than the
yield stress of fine-grained sintered material, then since in
general less plastic flow occurs in the sintered metal before the
fracture stress is attained, the UTS should bear some relation to
the yield strength and hence, unlike the dense metal, be affected
by the grain size.

Pore size is more difficult to measure than is grain size, and
its effect has been deduced from the observed phenomena, from
theoretical considerations, and with the aid of a two-dimensional
model.

As already observed, the grain size of sintered strip is related to the original particle size of the powder after a single roll-

Table III. Water-Atomized Copper Powders
Used for Determining the Effect of Grain Size
and Pore Size on Strength of Sintered Strip

Powder Designation	Percentage of Particles Passing through Sieve Openings of Sizes Shown				Average Particle Size
	%	μ	%	μ	μ
A	91	240–175	9	175	210
B	95	150–140	5	140	130
C	80	100–85	20	85	90
D	78	75–60	22	60	65
E	100	44	...	...	25

ing and sintering cycle. Moreover, it has been shown by Morgan [14] that the pore size in a green powder compact can be related to the particle size (see below). Consequently, powders of different particle sizes have been used in the preparation of strip with different pore sizes and grain sizes. The material used was the series of water-atomized powders with the size distributions shown in Table III. The powders were first deoxidized in hydrogen at 300°C.

Preliminary experiments had shown that all five powders could be compacted into coherent strip if an initial roll gap of −0.2 mm,* and a roll speed of 3-4 rpm were used. The strip produced under these conditions was sintered for 2 hr at 1005 ± 5°C in nitrogen. This provided specimens whose pore size and grain size both increased with increasing initial particle size.

In order to obtain specimens of substantially constant pore size but various grain sizes, sintered strip prepared from powder E was given a light reduction in cold rolling and annealed at different temperatures to promote grain growth.

Rectangular blanks, the approximate size of the tensile test piece, were cut from both series of sintered strip and their densities carefully measured. Tensile specimens were prepared from these blanks and annealed in the normal manner. The UTS

---

*A negative value denotes the presence of mutual roll pressure before rolling began.

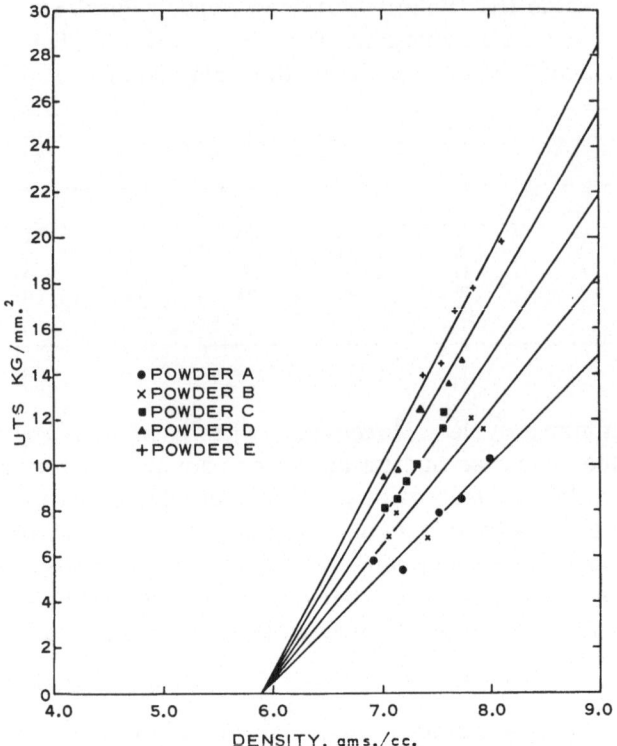

Fig. 6. UTS plotted against density of strip rolled from powders
A−E.

was measured and the broken test pieces mounted, polished, and etched, photographed, and the grain size determined from these photographs.

The sintered strip from powders A-E showed a progressive decrease in pore size and grain size with decreasing initial particle size, while the strip that had been cold-rolled and annealed to promote grain growth showed varying grain size but substantially constant pore size.

To obtain mean relative values of the strength at constant density, UTS density curves (Fig. 6) were plotted for specimens from powders A-E.

When UTS was plotted against grain size on a log/log scale
for the specimens from powders A-E, a linear relationship was
obtained with a slope of −0.49. Consequently, for both series of
specimens the UTS has been plotted against $D^{-1/2}$, where D is the
mean grain diameter (Fig. 7). The density of the second series of
specimens, of constant pore size, ranged from 7.9 to 8.1 g/cc,
and the UTS values of the A-E specimens plotted in Fig. 7 are
those obtaining at a density of 8.0 g/cc.

Extrapolation of the linear relationship of Fig. 7 to $D^{-1/2} = 0$,
i.e., the condition for a single crystal, shows a negative value of
UTS for specimens whose grain size and pore size are varying
(curve 1) and a value of 14.3 kg/mm^2 for the specimens of varying
grain size but constant pore size (curve 2). The validity of the
results shown in Fig. 7 is attested by the fact that the two lines in-
tersect at a point where the pores are of substantially equal size,
and obviously there is a common grain size at this point. Furthermore, the UTS (14.3 kg/mm^2) obtained by extrapolating curve 2
of Fig. 7 to $D^{-1/2} = 0$, lies within the range of values (12.9-35.0
kg/mm^2), depending on orientation, given by Czochralski [15] for
a single crystal.

The difference in the slopes of curves 1 and 2 of Fig. 7, and
the meaningless of a "negative" UTS, show that there is a factor,
other than grain size, whose influence in reducing the UTS increases as the grain size is increased. The difference between
the two series of specimens is one of pore size, so that the difference in behavior may be ascribed to this cause. Thus, if pore
size is so important a factor, it would seem desirable to inquire
how different pore sizes − at constant total porosity − arise in the
first place.

It has been shown by Morgan [14] that the average circular in-
scribed pore diameter for the most probable packing of any-size
grading of spheres between diameters d and d/2 is 0.193d. In
Fig. 8 (curve A) the UTS of specimens A-E at a density of 8.0 g
per cc is plotted against 0.193d, where d is the average initial
diameter of the particles in each powder (Table III). Although
Morgan's relationship is for unsintered spheres, the effect of sintering is to spheroidize the pores, so that to a first approximation
the relationship between original particle size and pore size may
be expected to be valid for sintered material.

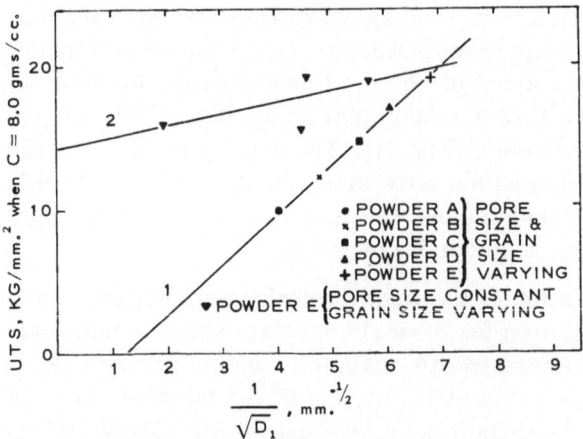

Fig. 7. UTS (at a density of 8.0 g/cc) plotted against the
inverse square root of the mean grain diameter.

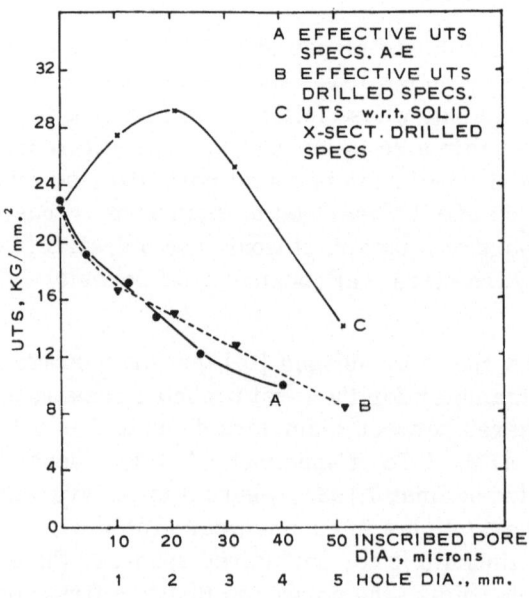

Fig. 8. UTS plotted against diameter of voids.

In order to clarify the effect of pore size on the strength of material with constant grain size, the following model was devised. Five tensile specimens were prepared from extruded copper strip having a cross section of 3.2 × 31.8 mm ($\frac{1}{8}$ in. × $1\frac{1}{4}$ in.). Holes were drilled in four of them. The diameter, number, and position of the holes were so chosen that in each specimen the total "porosity" was constant, the minimum cross section of the metal was constant, and the ratio of hole diameter to the distance between hole centers was approximately constant (Fig. 9).*

The five specimens were annealed after preparation and broken in tension. Hole diameter is plotted against effective UTS in Fig. 8 (curve B). It can be seen that the form of the curve follows closely that obtained for specimens A-E (curve A). When the UTS of the drilled specimens is calculated with respect to the residual solid cross section, the curve of strength against hole diameter shows a maximum (curve C). This indicates an increased strength per unit area of solid cross section, presumably caused by stress-interference effects in the regions surrounding the holes. This is in accord with Thum's finding that stress concentration can be substantially relieved by a series of slits placed at regular intervals [16].

It is obvious that the stress effects in the drilled specimens will only approximate to those engendered by closed pores, since the former show approximately two-dimensional porosity.

From the strength/pore-size relationship of Fig. 8 (curve A) and the strength/hole-size relationship (curve B), it is clear that the effective UTS (σ') for a specimen of given grain size and total porosity is related to the pore diameter P in the form

$$\sigma' \propto P^{-n}$$

and the curves of Fig. 7 may be re-examined in this light with reference to Figs. 10 and 11.

(i) With reference to Fig. 10: The experimental curves in Fig. 10 (a) are the curves of Fig. 7 with the $D^{-1/2}$ axis

---

*An improvement would be to space the holes in a close-packed hexagonal pattern in each specimen.

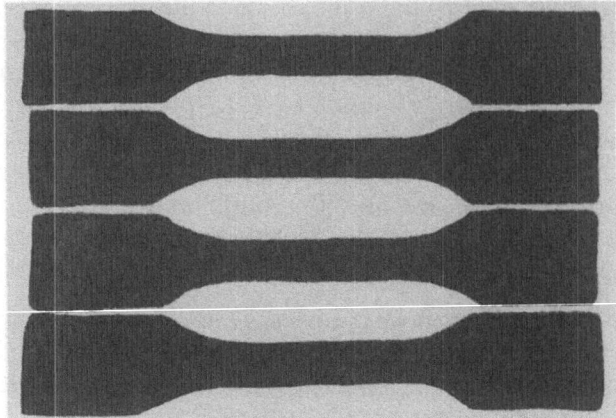

Fig. 9. Drilled specimens. $\times^2/_5$.

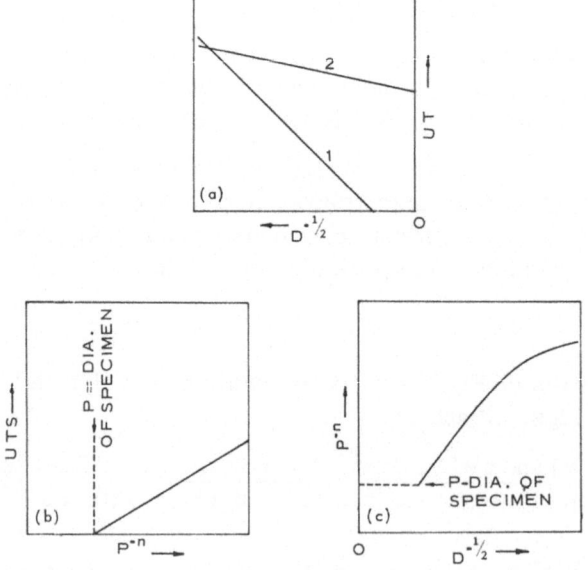

Fig. 10. Curves illustrating the effect of pore size and grain size
(see text).

reversed. Curve 1 shows $UTS/D^{-1/2}$ for specimens whose grain size and pore size are varying. Curve 2 shows $UTS/D^{-1/2}$ for specimens of varying grain size but constant pore size. Thus, there is a variable factor present in curve 1 which is not taken into account except indirectly by observation of its effect on the UTS.

For specimens of constant grain size and a given total porosity, the strength decreases as the pore size is increased, as shown schematically in Fig. 10 (b). The strength must be zero when the pore diameter becomes equal to the breadth or thickness of the parallel gauge-length of the specimen.

For powders of different initial particle sizes there is a unique relationship between grain size and pore size for a given rolling and sintering treatment, and a given total porosity. This is shown schematically in Fig. 10 (c). By subsequent treatment, the grain size can be varied while the pore size remains substantially constant, yielding a series of curves between that shown in Fig. 10 (c) and the P^{-n} axis.

By combining curve 2 of Fig. 10 (a) with the curves of Fig. 10 (b), and 10 (c), the relationship between UTS, grain size, and pore size shown in Fig. 11 is obtained.

(ii) With reference to Fig. 11: The limiting surface SRT exists by virtue of the original $P^{-n}/D^{-1/2}$ relationship of Fig. 10 (c), i.e., it is determined by the initial treatment.

It can be seen that curve 1 of Fig. 7, representing the variation of UTS with grain size and pore size, should in fact be plotted in three dimensions as a line such as KL in Fig. 11. Curve 1 in Fig. 7 is, in effect, the projection of KL onto the $UTS/D^{-1/2}$ plane.

It is clear that for specimens of a given total porosity, the main factor determining their strength is pore size rather than grain size.

IV. Electrical-Resistance Measurements

1. Method of Study

The electrical resistance of green or sintered strip depends upon the following factors: (a) the inherent resistance of the metal concerned, (b) the porosity, and (c) the resistance at the particle

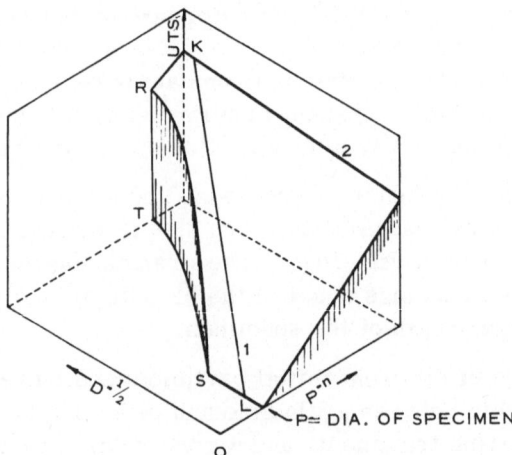

Fig. 11. Effect of pore size and grain size
on UTS of sintered material (schematic).

interfaces caused by microporosity and the presence of impurity
films, e.g., oxide.   Factor (a) is constant for a given metal, but
(b) and (c) will vary with the pressing and sintering conditions.
Measurements have been made of the variation of resistance with
density in green and sintered strips, and the effect of different
sintering times at constant temperature has also been studied. The
resistance present at the particle interfaces in green strip will depend to a considerable extent upon the relative movement that
takes place at an interface during compaction, as this will determine to what extent surface films are removed, and metal/metal
contacts set up between the particles.   This being so, it was considered that measurements of the variation of resistance with direction in rolled strip should be made, to determine whether there
is any marked difference in the relative movement between particles in the rolling direction as compared with the transverse direction.   In order to estimate the influence of this factor, the resistance measurements must be made in such a way as to eliminate any influence of variations in porosity.   There will be no anisotropic effects on the electrical resistance inherently due to the
metal, since copper having a cubic structure has an isotropic electrical resistance.

A Kelvin bridge with a range from 0.1 $\mu\Omega$ to 1.1 Ω was used for making resistance measurements between needle probes in contact with the specimen. The errors arising from contact resistances and thermoelectric effects are relatively small compared with the error that may be introduced by assuming that the material is homogeneous and computing its resistance from its cross section and the separation of the probes. The need to allow for inhomogeneities in the specimen has a considerable bearing on the measurement of the resistance of green strip and is dealt with at some length elsewhere [7]. In brief, it is necessary to take the mean of a number of readings with the probes in different positions.

The first methods used to determine whether there was any difference between the resistance of green sheet parallel to the rolling direction and that in the transverse direction involved making measurements about a fixed point in the specimen. However, the general pattern of density distribution affects the distribution of the flow lines and equipotential lines in the specimen, and the resistivity cannot be found from a single measurement about the chosen point if the size of the test specimen is at all appreciable, as this increases the possibility of density variations.

Since the object of the experiment was to determine whether any difference exists in interparticle bonding in the parallel and transverse directions, it is sufficient to measure the resistance along different lengths about the point of intersection of two straight lines parallel to these two directions. Provided that a square specimen is employed and a sufficient number of measurements made with the probes at different distances apart, the relative values of the resistances obtained by extrapolation to very small distances should give a valid indication of the relative degree of bonding in the two directions. Since density is meaningless at a point, this result is independent of porosity variations and is not affected by current and potential distribution in the rest of the sheet. It must be emphasized that in each case E/I, the ratio of the potential difference between the probes to the current flowing between them, is being measured and extrapolated to a value of E/I for very small distances. The relation which this bears to the average resistivity of the material will depend on the distribution of density in the specimen.

Table IV. Electrolytic Copper Powders
Used for Resistance Measurements

Sieve No	Size, μ	Batch 1	Batch 2	Batch 3
		wt.-%	wt.-%	wt.-%
+100 mesh	147	Trace	0·1	0·2
−100 +200	147–74	3·8	19·8	21·6
−200 +325	74–44	20·0	26·1	36·7
−325	≦44	76·2	54·0	41·5
Sub-sieve range: <44				
Specific surface cm.²/g.		470	479	520
Apparent density g./c.c.		2·46	2·72	2·85
Cu		98·80	99·81	
Zn		Nil	Nil	
Fe		0·13	Trace	
Ni		Nil	Nil	Not
Sn		0·06	Nil	determined
Pb		0·26	Nil	
SiO₂		0·14	0·06	
O₂		0·59	0·06	

2.   Experimental

Electrolytic copper powder, from batches 1, 2, and 3 (Table IV) was used in all the experiments.

For studying the variation of resistance with density, all measurements were made in the rolling direction on strips 12 × 2.0 × 0.06 cm, cut from the green material. The sintered material had been heated for 2 hr at 1000°C in pure nitrogen. To study the effect of sintering time, strips were cut as before from green material and sintered at 1000°C for times up to 120 min, the same specimen being used for each of the sintering times. For the above experiments the measuring probes were loaded with a 100-g weight.

For the measurements parallel to and across the rolling direction, square specimens were used, measuring 10 × 10 × 0.06 cm, cut from the center of green strip 19 cm wide. The probe separation was varied from approximately 7 to 0.3 cm. In making measurements, one probe was kept a fixed distance from one current contact of the bridge and the other probe moved to various distances along the line of current flow. The specimen was then rotated through 90° and the measurements repeated about the same point. This technique is necessary to ensure the maximum accuracy in the results.

Deoxidized powder was prepared by heating in pure hydrogen
at 330°C for 2 hr. Oxidized powder was prepared by heating powder for several hours at 100°C in air.

3. Results and Discussion

(a) Effect of Density and Sintering Time

Electrical conductivity is plotted against percentage porosity
in Fig. 12 for green strip rolled from powder of Batch 1. When
this relationship is extrapolated to zero conductivity, it is seen
that this occurs at 33% porosity, i.e., at a density of 5.96 g/cc.
But green strip having a density of only 5.5 g/cc is just coherent
and must have a conductivity greater than zero. Hence, the linear
relationship must be invalid at high porosities. Extrapolation to
zero porosity gives a value of conductivity of 0.027 cm/$\mu\Omega$ (a resistivity of 37 $\mu\Omega$/cm). The main impurities in this batch of copper powder were lead (0.26%) and iron (0.13%). The presence of
as much as 1% lead in copper decreases its conductivity only to
98% of that for the pure metal [17]. Iron has a marked effect on
the conductivity of copper unless oxygen is present to remove it
from solid solution [18]. However, the resistance of green material will be mainly dependent on the interparticle bonds and
hence the form in which the oxide is present is likely to play a
dominant part; it would seem probable, therefore, that the reason
for the relatively low conductivity of high-density green strip lies
mainly in the nature of its oxide dispersion. A further contribu-
tory factor might be the work-hardened state of the green strip,
but this probably plays a minor role, since the increase in resistivity of cold-deformed fused metal is not more than 5% [19].

The variation of resistivity with sintering time lends weight
to the preceding argument. The mean results for twelve specimens, all of comparable density, are shown in Fig. 13. Likhtman
and Nazarov [20], plotting resistivity against sintering temperature for electrolytic copper powder, found a rapid decrease in resistivity in the range 100-500°C. They attributed this to the reduction of the oxide films and a consequent change in the charac-
ter of the interparticle contacts. In the time range 0-5 min (Fig.
13), a similar rapid decrease is shown, the greater part of which
(from 214.0 to 3.0 ω/cm × 10^{-6}) would seem to take place in a sintering period of considerably less than 5 min. It probably occurs

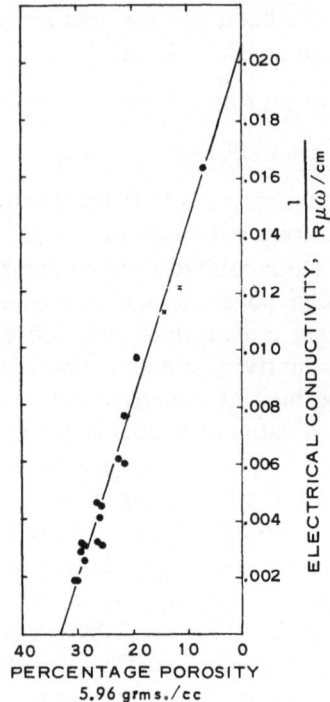

Fig. 12. Variation of conductivity with
porosity for green strip.

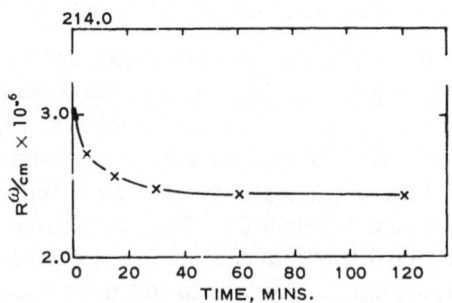

Fig. 13. Resistance plotted against sintering time
at 1015°C.

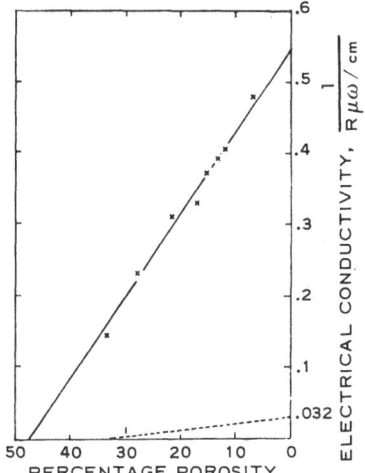

Fig. 14. Conductivity plotted against
porosity for sintered strip.

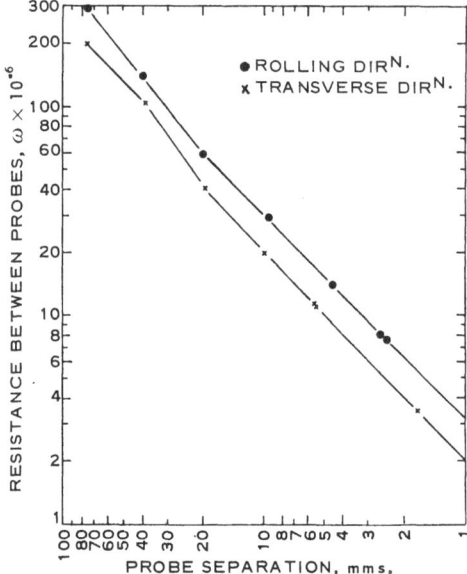

Fig. 15. Resistance in the rolling and transverse
directions for different probe separations: deoxidized powder.

for the same reason as that suggested by the Russian authors,
since deliberately oxidized specimens emerged bright after sintering under the conditions described. Complete reduction of the
oxide need not be postulated, however; it is only necessary that
the remaining oxide form discrete particles.

The relation between conductivity and porosity is shown in
Fig. 14 for strip rolled from Batch 2 powder to different densities
and sintered in nitrogen for 125 min at 1005°±2°C. Figures 14 also shows, as a dotted line, the conductivity/porosity relationship
of Fig. 12. The extrapolation of a mean line through the experimental points gives a value of 0.548 cm/$\mu\Omega$ for the conductivity
at zero porosity. This is about 93% of 0.599 cm/$\mu\Omega$, the conductivity quoted for pure copper. The corresponding resistivities
are 1.82 and 1.67 $\mu\Omega$/cm. The lower conductivity of the material
prepared from powder may be attributed to the trace of iron present. An iron content as low as 0.05 wt.% can reduce the conductivity to 78% of its value for pure copper if no oxygen is present
to remove the iron from solid solution [18], and the initial oxide
content of the powder of Batch 2 was very low (Table IV).

When the mean line of Fig. 14 is extrapolated to zero conductivity, the corresponding porosity is 46.5%. The maximum porosity that can be achieved by systematically packing equal-sized
spheres is 47.6%; it is clear that in such a system a greater porosity would be accompanied by loss of contact between the spheres
and zero conductivity. It is not clear why this should be so in a
system of nonspherical particles of different sizes. Grootenhuis
et al. [21] suggest that this affords an insight into the mechanism
of sintering, since the grain growth between the particles at the
points of contact produces a system whose density distribution is
independent of the shape of the original particles.

The existence of a linear relationship between conductivity
and percentage porosity for both green and sintered material
merely indicates that the number of metal–metal bonds increases
with the density. It does not necessarily mean that "the resistance
is mainly confined to the particles and not to the interparticle
bonds," which is the view advanced by Grootenhuis and his co-
workers [21] on this relationship for sintered materials.

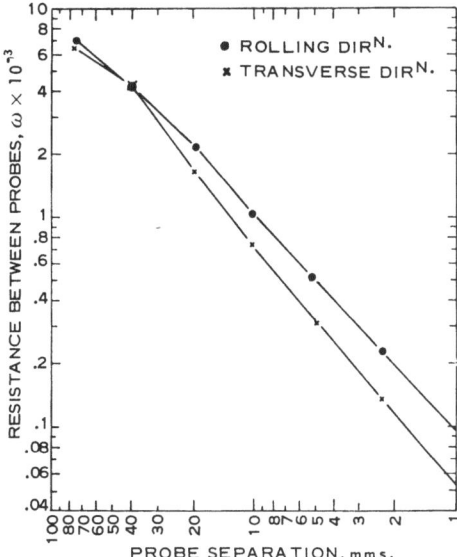

Fig. 16. Resistance in the rolling and transverse directions for different probe separations: oxidized powder.

(b) Variation of Resistance with Direction

Typical results for these experiments are shown in Fig. 15 and 16, where $\log R'$, the resistance measured between the probes, is plotted against $\log l$, the separation of the probes. It can be seen that a linear relationship exists for values of l less than approximately 2.0 cm. This is interpreted that there exist only small variations in density over these regions. At higher values of l, the linear relationship is not maintained. This is attributed to variations in density.

The variations with probe separation of the resistance in the rolling and transverse directions for sheet rolled from deoxidized and heavily oxidized powder (Batch 3, Table IV) are shown in Figs. 15 and 16, respectively. The specimens were of the same dimensions to within $\pm \frac{1}{2}\%$ and the difference in the resistance for any given probe separation, amounting as it does to about 15 times, must be due to the different oxide contents. When the probe separation is extrapolated to 1 mm, the resistance in the rolling direction is 60% higher than that in the transverse direction for deoxidized powder and 88% higher for heavily oxidized powder. This is in accord with the view that the difference between the resistance in the two directions arises from the mode of deformation of the particles, whereby there is relative movement of the particles in the rolling direction in the early stages of compaction. This would tend to remove surface films from those parts of the particles parallel to the rolling direction more than from those in the transverse direction, thus giving a greater number of metal–metal bonds per unit area. When the particles are surrounded by a heavy oxide layer, the effect should be more pronounced.

References

1. P. E. Evans and G. C. Smith, "Symposium on Powder Metallurgy 1954" (Special Rept. No. 58), London, Iron and Steel Institute (1956), p. 131.
2. P. E. Evans and G. C. Smith, Sheet Metal Ind. 32 : 589 (1955).
3. W. M. Baldwin, Trans. Am. Inst. Min. Met. Eng. 166 : 591 (1946).
4. R. M. Brick and M. A. Williamson, ibid., 143 : 84 (1941).
5. H. Hu, P. R. Sperry, and P. A. Beck, J. Metals, 4 : 76 (1952).
6. F. von Göler and G. Sachs, Z. Physik, 41 : 843, 889 (1927); 56 : 477, 485 (1929).
7. P. E. Evans, Ph. D. dissertation. Cambridge University (1956).
8. E. Scheuer, Metallwirtschaft, 14 : 365 (1935).
9. C. G. Goetzel, A Treatise on Powder Metallurgy, Vol. I, New York, Interscience Publishers (1949), p. 88.
10. C. G. Goetzel, op. cit., Vol. II, New York, Interscience Publishers (1950), p. 504.
11. W. T. Pell-Wallpole, Foundry Trade J., 98 : 341 (1955).
12. O. J. Dunmore, Ph. D. dissertation, Cambridge University (1955).
13. R. P. Carreker and W. R. Hibbard, Acta Met., 1 : 654 (1953).
14. V. T. Morgan, "Symposium on Powder Metallurgy 1954" (Special Rept. No. 58), London, Iron and Steel Institute (1956), p. 81.
15. J. Czochralski, Proc. Internatl. Congr. Appl. Mechanics, Delft (1924), p. 67.
16. A. Thum and H. Ude, Giesserei, 16 : 501 (1929).
17. R. A. Wilkins and E. S. Bunn, "Copper and Copper-Base Alloys," New York and London, McGraw-Hill (1943), p. 21.

18. J. S. Smart and A. A. Smith, Trans. Am. Inst. Min. Met. Eng. 147:48 (1942).

19. C. G. Goetzel, op. cit., Vol. I, p. 541.

20. V. I. Likhtman and L. T. Nazarov, Dokl. Akad. Nauk SSSR, 78:749 (1951).

21. P. Grootenhuis, R. W. Powell, and R. P. Tye, Proc. Phys. Soc., 65B:502 (1952).

An Examination of the Compaction Process

The compaction process is examined in detail. It is shown that, where particle deformation is concerned, compaction by rolling is similar to compaction by static pressing, with the addition of elongation of the particles in the rolling direction when the rolling pressure is sufficiently high. A method for determining the average roll pressure is described. A comparison of the rolling of a metal powder with the rolling of a solid bar, and the determination of the effect of particle shape and mean size, indicates that not only roll/powder friction but also the slip between particles plays an important role in the compaction process. This leads to an examination of the flow properties of powders, which are measured in terms of a "powder-viscosity factor" that indicates whether and at what order of rolling speed a powder can be coherently compacted. Finally, a mechanism of compaction is proposed on the basis of the present findings and on the authors' earlier work.

I. Introduction

The authors have discussed their results regarding some of the properties of strip rolled from copper powder, with brief reference to the findings of other workers, in Part II of this paper (p. 121), and elsewhere. A more detailed examination of the process whereby metal powders are compacted by rolling is now presented. Although a standard rolling mill has been used, it is thought that the conclusions reached are, in general, valid for powder compaction in a mill whose rolls are set in the same horizontal plane.

II. Deformation of Particles in Pressing

and in Rolling

Substantially spherical lead shot, about 1.5 mm in diameter, was initially used to compare the deformation in pressing with that in rolling. The shot was pressed in a $1\frac{1}{2}$-in.-diameter steel die and rolled in copper sheaths. The pressed shot showed a fairly regular close packing (Fig. 1), and the rolled shot showed a similar structure with elongation in the rolling direction superimposed (Fig. 2).

The stages in the compaction of copper powder by rolling were then examined directly, by mounting the wedge of coherent material available when the rolls were stopped before all the powder had passed through. The "taper section" so obtained showed the effect of increasing roll pressure from the minimum pressure necessary for coherency up to the maximum pressure dictated by the initial setting of the roll gap. It is clear that the metal particles are first pressed into intimate contact and then, if the pressure is sufficiently high, elongated in the rolling direction. The degree of elongation depends on the pressure and on the ductility of the metal. Copper powders may undergo a high degree of deformation, as can be seen from Fig. 3, which is made up from photomicrographs in the three principal directions. Compression has occurred in the rolling plane and elongation in the rolling direction. Localized deformation of the particles may also take place (Fig. 4); the particles have here been forced apart during polishing.

III. The Specific Roll Pressure

Orowan has proposed a theory [1] that enables the pressure distribution between the rolls and the material to be obtained when the latter shows variable resistance to plane homogeneous compression. A theory applicable to the rolling of metal powders might conceivably be evolved by a similar approach to the problem.

The variation in "yield stress" of a mass of powder increases the difficulty of determining the specific roll pressure by any of the methods that have been used for dense metals. An exception is that of Smith, Scott, and Sylwestrowicz [2], which employs a pin set in one roll and bearing on a piezoelectric pressure-measuring cell. It shows that there is a considerable variation in specific roll pressure over the arc of contact. The special apparatus involved has precluded its use in the present experiments. It is pointed out in their paper that even the mean roll pressure cannot be calculated from a knowledge of roll-load measurements alone. However, the method used by Wallquist [3] ignores this, and measurements of total roll load were made and the mean roll pressure calculated from them. Cook and Parker [4] and Sims [5] give details of methods used to compute total roll loads. Both these methods require a knowledge of the resistance of the material to plane homogeneous compression. For powder rolling,

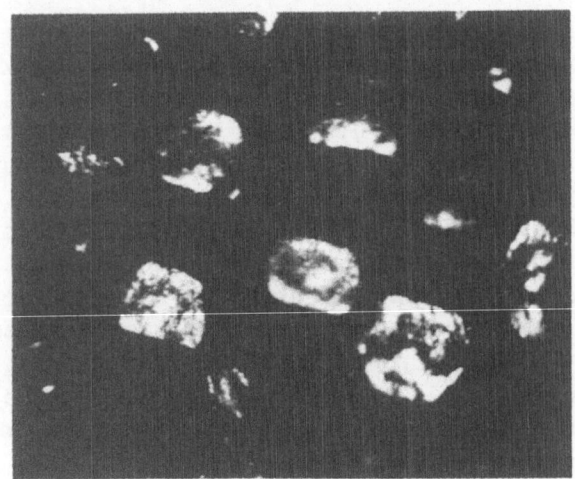

Fig. 1. Bed of lead shot compressed in a die. × 18.

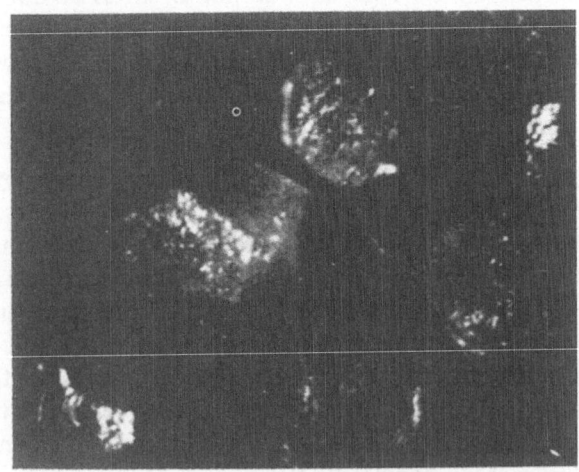

Fig. 2. Bed of rolled lead shot. × 25.

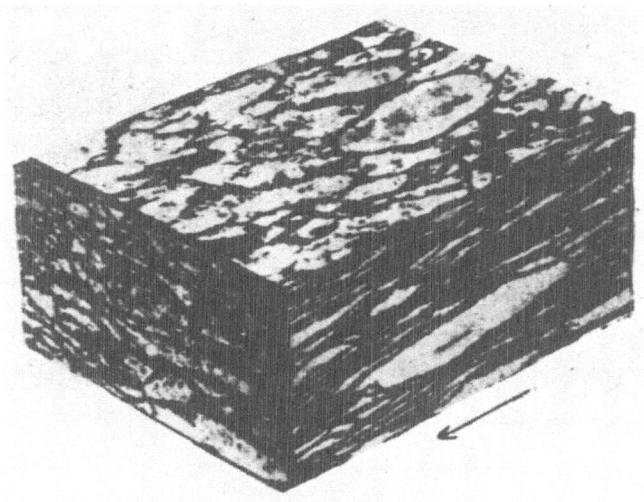

Fig. 3. Illustrating deformation of particles in "green" strip made from copper powder. Model made up from photomicrographs (x 250) in the three principal directions. Arrow shows rolling direction.

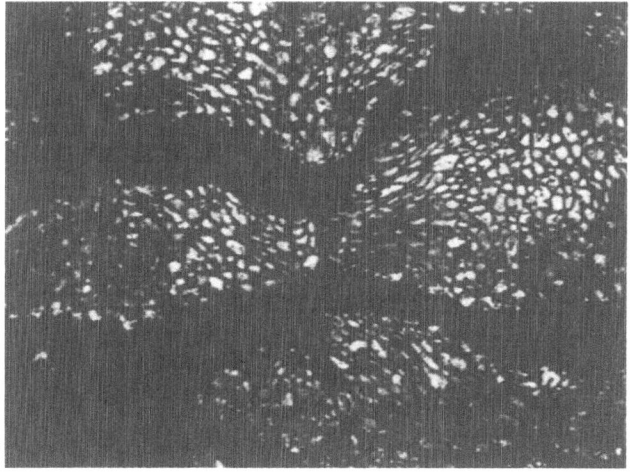

Fig. 4. Localized deformation of particles in strip. Particles originally in contact have been forced apart during polishing. x 250.

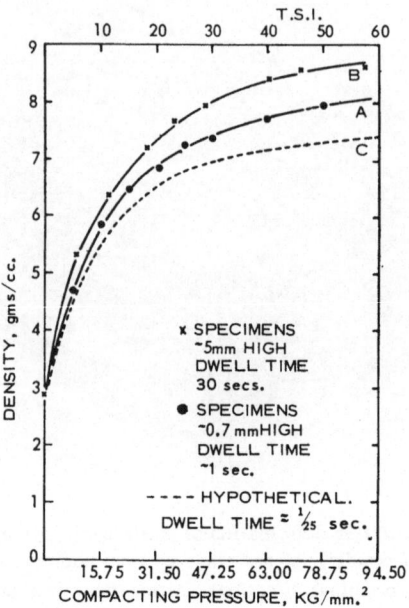

Fig. 5. Pressure/density relationship for
powder compacts made from electrolytic
copper powder.

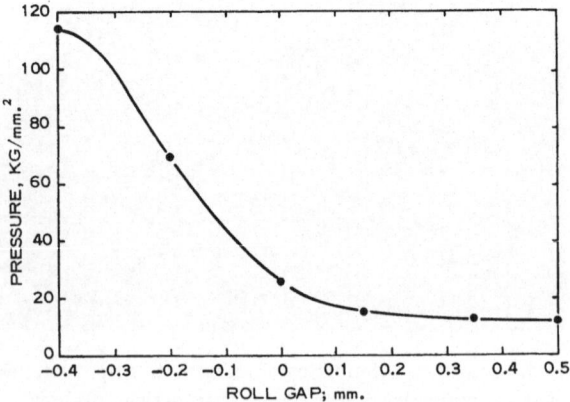

Fig. 6. Variation of pressure with initial roll gap. Negative
values denote presence of mutual roll pressure before powder
rolling began.

however, the varying yield stress and the difficulty of determining the area of contact were thought to justify the adoption of the
following method in preference to those described above.

Small quantities of electrolytic copper powder were pressed
in a double-action cylindrical die under different pressures to
yield compacts 7.96 mm in diameter and varying in thickness from
0.54 to 1.12 mm. No lubricant was used. The density of each
compact was determined by weighing and measuring. This enabled
a pressure/density relationship to be plotted (Fig. 5, curve A).
Density measurements on strip rolled from the same powder at
different settings of the roll gap (yielding strip 0.3-1.0 mm thick)
allow the approximate mean rolling pressure to be obtained for
any initial setting of the roll gap (Fig. 6).

The measures taken to minimize the difference in the two
modes of compaction were: (i) to limit the thickness of the compacts to the approximate range of thickness obtained in rolling;
and, (ii) to raise the pressure on the plungers at a high rate to a
maximum value and then to decrease it to zero as quickly as possible. The pressures plotted in Fig. 5 (curve A) are the maximum
pressures, obtained in this way. *

The effect on density of varying the time during which pressure is applied is shown by the experimental results obtained when
powder from the same batch as that previously used was pressed
in a die under loads that were allowed to "dwell" for 30 sec. These
compacts were about 5 mm high and of the same diameter as those
previously described; the pressure/density relationship is shown
in Fig. 5 (curve B). The greater height of these compacts would
be expected to decrease the average density under given pressing
conditions, and hence the increased density obtained for a given
pressure must be due to the increased dwell-time of the maximum
load. The strip whose density was measured to determine the
pressure/roll-gap relationship of Fig. 6 was rolled at a speed of
24 rpm, the roll radius was 10.1 cm and, for this powder, the
gripping angle was about 6°; hence, the maximum time during
which any appreciable pressure was applied was of the order of

---

* A possible improvement would be to replace the die by a Plasticine mold, thereby
eliminating the wall effect.

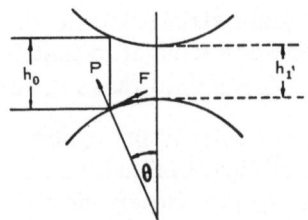

Fig. 7. Schematic representation of
entry of solid bar between rolls.

$\frac{1}{25}$ sec. The time of application of maximum pressure cannot be
determined, since the pressure distribution over the arc of gripping contact is not known, but it must be less than the $\frac{1}{25}$ sec
quoted above. If it is assumed that the dwell-time for the thin
compacts (Fig. 5, curve A) was of the order of 1 sec, the approximate effect of the reduced dwell-time associated with a rolling
speed of 24 rpm, i.e., $\frac{1}{25}$ sec, is probably similar to that shown
in the dotted line (Fig. 5, curve C). The experimentally obtained
curve A of Fig. 5 has been used, however, in calculating the roll
pressures shown in Fig. 6.

IV. Powder Rolling and Dense Metal Rolling

It is clear from the experimental evidence presented above
that the powder, once it has been fed to the rolls, is subjected to
an increasing degree of compaction up to the maximum dictated
by the setting of the roll gap.

Initially, powder is carried from the bottom of the chute toward the roll gap as a relatively uncompacted mass. With increasing constriction, however, the particles are forced into more intimate contact, with the possibility of a rapid increase in the number of direct metal-to-metal bonds. From here onward the material behaves in its passage between the rolls in a fashion intermediate between that of the loose powder and that of a solid metal
bar.

The entry of a solid bar between the rolls becomes impossible when the horizontal forces are zero (Fig. 7).

Table I. Gripping Angles
for Various Types of Powder

Powder (all −100 mesh)	Gripping Angle, ±$\frac{1}{4}$°
Electrolytic	6°
Water-atomized	3$\frac{1}{2}$°
Air-atomized	1$\frac{1}{2}$°

Then

$$P \sin \theta = F \cos \theta$$

i.e.,

$$F/P = \tan \theta$$

But $F/P = \mu \tan f$, where f is the angle of friction between the bar and the rolls. Thus, the bar cannot be drawn into the rolls when the contact angle θ exceeds the friction angle f .

Hence, an approximate value for the angle of friction f may be found [6]. A bar of initial thickness h_0 is held gently in contact with the rolls and the roll gap is slowly increased from zero to the point at which the metal is just.drawn between the rolls. Let the rolled thickness be h_1 (Fig. 7); then

$$\frac{h_0 - h_1}{2} = R(1 - \cos f) \tag{1}$$

where R is the roll diameter.

For annealed solid copper rolled between smooth steel rolls, the following experimental values were obtained: $R = 101.5$, $h_0 = 5.28$, and $h_1 = 2.40$ mm. After substituting in (1), $f = 9°45'$ and $\mu = 0.17$. The maximum draft $(h_0 - h_1)$ occurs when the contact angle θ is equal to the angle of friction f.

For a given electrolytic copper powder the apparent density was 2.4 g/cc and the lowest density of coherent green strip was approximately 5.5 g/cc. Thus, a compaction ratio h_0/h_1 of at least 5.5/2.4 is necessary to roll this powder into coherent strip. If it is assumed: (a) that the same value of f obtains for the powder as for the solid bar; (b) that no slipping occurs between the powder particles; and, (c) that the pressure distribution and μ remain constant over the arc of contact, it is possible to calculate

the theoretical maximum thickness of strip that could be rolled
from the powder. For the powder in question, from equation (1),
we have

$$\frac{h_0}{h_1} = \frac{2R(1 - \cos f)}{h_1} + 1 = \frac{5.5}{2.4} \qquad (2)$$

Hence,

$$h_1 = 2.2 \text{ mm}$$

It was found by experiment that this powder yielded a strip
0.8 mm thick at a density of 5.5 g/cc. Substituting this value of
h_1 in equation (2) gives f = 5°48' = 6°, approximately.

The discrepancy between this value of f and that obtained for
the solid bar clearly indicates that the assumptions made above
are not valid. If the variations in pressure and μ over the arc of
contact are ignored, being in this context of secondary importance,
it means that, during rolling, slip occurs between the particles at
least up to that point on the roll surface for which θ is equal to α,
the "gripping angle" for the powder.

The value of α may be obtained directly by stopping the rolls
before all the powder has been compacted and gently blowing away
the uncompacted powder. Values of α (initial roll gap 0 mm) are
given in Table I for various types of powder.

It can be seen that the gripping angles for the powders differ
markedly from that for a solid bar. This, and the variation in
gripping angle for powders of differing general shapes, indicates
that slipping between the powder particles does play an important
part in compaction by rolling. The state of affairs in the powder
in the roll gap can be represented as in Fig. 8, where zone 1 contains the powder from the bottom of the chute to the point where
it begins to be gripped by the rolls, and zone 2 contains the powder between this point and the line joining the roll centers. It has
also been observed that whereas the gripping angle is, within the
limits of determination, independent of the particle size for powders whose average size ranges from 200 to 25 μ, the strip thickness decreases with particle size. As the average particle size
decreases, the surface area per unit mass of powder increases,
and this obviously affects the thickness of the strip produced under
fixed rolling conditions. Hence, strip thickness is controlled both

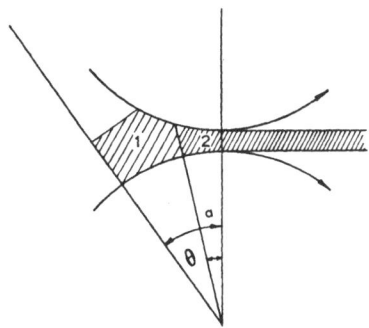

Fig. 8. Schematic representation of state
of powder in the roll gap. Zone 1, incoherent powder; zone 2, coherent powder.

by the gripping angle and by the average particle size. The very
small gripping angle observed for air-atomized powder is interesting, in that it has not been found possible to roll a coherent
strip from air-atomized powder, although such a powder can just
be compacted in a die. It might be possible to compact such powder by rolling if rolls of sufficiently large diameter were used,
thereby increasing the compaction ratio.

V. Roll Powder Friction

1. General

The variation in the gripping angle with particle shape (Table
I), may be attributed to roll/particle friction, since particles of
the same general shape but different mean particle sizes showed
the same gripping angle. Interparticle friction, on the other hand,
and the inability of a loose powder to withstand a shearing stress
of any magnitude constitute the main differences between rolling
powders and rolling a solid bar. Furthermore, even when the
powder has passed the point of gripping, its yield stress must
vary, increasing as the bonding between particles increases and
as work-hardening proceeds. The increased bonding reduces the
possibility of relative particle movement.

Before considering the effect of the flow properties of the powder, it is worth examining the part played in the compaction process by roll/powder friction.

2.   Factors Affecting Roll/Powder Friction

If the roll/powder friction is increased, the gripping angle α becomes greater, thus increasing the thickness of the powder gripped by the rolls. Hence, for a given compaction ratio, i.e., a given initial and final density, the thickness of the strip is increased.

By analogy with the rolling of dense metals, the factors most likely to affect the roll/particle friction are (a) the surface condition of the rolls; (b) the temperature of rolling; (c) pressure feeding of material to the rolls; and, (d) the roll speed.

(a) Surface Condition of Rolls

Three types of surface finish (shot-blasted, smooth matte, and bright-polished) have been used to compact powder into strip. Shot-blasted rolls yielded a strip 40% thicker than that produced by smooth matte rolls, and 100% thicker than the strip produced by bright-polished rolls for a given powder feed at a constant setting of roll speed and roll gap. This result is in agreement with the effect of surface condition on the coefficient of friction during the rolling of solid metals [7].

(b) Rolling Temperature

Electrolytic copper powder has been rolled at 300°C, using a modified form of feed apparatus that enabled the powder to be heated in a controlled atmosphere before it reached the roll gap. The density of the green strip was increased by about 3% and the thickness by about 25%, compared with the values obtained for the same powder and rolling conditions at room temperature. No data are available for solid copper, but in the case of steel the friction increases with temperature up to almost 400°C.

(c) Forced Feed

Since it is known that "bumping" a bar into the rolls can increase the frictional forces between the bar and the rolls, enabling it to be drawn into the roll gap [8], an attempt has been made to feed powder under pressure to the rolls.

The apparatus (Fig. 9) consists of a horizontal tube, one end of which is shaped to fit the roll gap. The other end carries a ball bearing and a thrust bearing locating the shaft of an Archi-

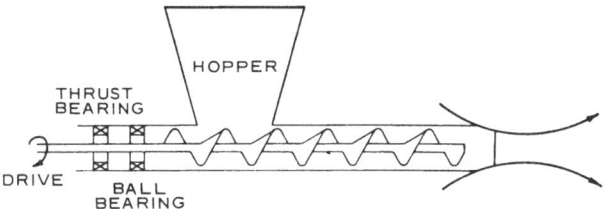

Fig. 9. Apparatus for forced feed of powder to rolls (schematic).

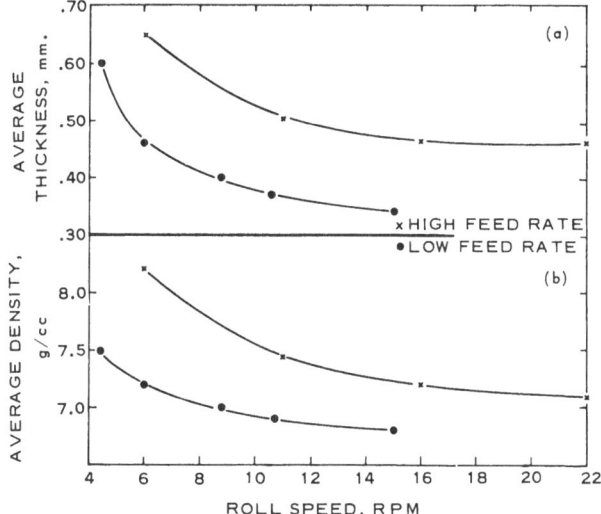

Fig. 10. Effect of roll speed and external powder feed on
thickness and density of green strip.

medean screw which feeds powder forward into the roll gap. When
the roll speed is adjusted to the feed rate, powder can be compacted into strip, but no appreciable increase in thickness or density has been obtained by this method. The increase in pressure
is very small, and if the feed rate is increased, the powder mass
shears at the roll end of the screw and at the point of entry of the
material from the hopper. A cylinder of powder is then carried
around by the screw with no forward feed.

It is possible that a forced-feed technique may be more effective at high rolling speeds where the supply of powder by a

gravity feed can become insufficient to allow a coherent sheet to be produced, or for powders with poor flow properties.

(d) Rolling Speed and Powder Feed

For dense metals the coefficient of friction decreases as the roll speed increases [9]. There is no direct parallel in the rolling of metal powders, since, as the rolling speed is increased, the relative feed rate, at a constant nominal feed, decreases. Hence, the effect of rolling speed has to be considered in conjunction with the powder feed, whereas, for a solid bar, the feed rate always keeps pace with the rolling speed. For this reason, the roll/powder friction is of secondary importance to variations in feed rate and speed.

In compaction by rolling there are three rate-controlling elements: (1) the hopper aperture, which determines the rate at which a given powder reaches the rolls; (2) the roll speed and roll gap, which determine the rate at which a given powder is compacted into strip; and, (3) the flow properties of the powder which, for a given roll speed and gap, determine the rate at which the powder passes from the incoherent to the coherent zone. It is convenient to define two feed rates: (i) the "external" feed, which is determined by the aperture of the hopper and the flow properties of the powder; and, (ii) the "internal" feed, which is determined by roll speed and gap and by the flow properties of the powder. The superincumbent weight of powder in zone 1 (Fig. 8) depends, in an obvious manner, on the relationship between these three elements.

For a low, constant external feed rate of a certain electrolytic copper powder and with a fixed roll gap, the thickness and density of the strip decreased as the roll speed was increased (Fig. 10). When the external rate was increased fourfold, the thickness and density of the strip were greater than the values obtained for the lower feed rate at any given roll speed (Fig. 10). The powder supply became inadequate, with the low feed rate, at speeds above ~16 rpm, and with the higher feed rate above ~25 rpm. At higher speeds powder passed through the rolls without being coherently compacted.

The increase in thickness and density with increased feed, at constant roll speed, is probably due to an increase in density of the loose powder as it moves through zone 1. At the bottom of the chute the density of the powder must be approximately equal to its apparent density. At any given roll speed there will be a greater chance of a high packing density in zone 1 as the volume of powder in that zone is increased.

Another factor that probably influences the packing density of the powder entering the rolls is the expulsion of entrapped air from the powder as it is fed into the roll gap. Cine films taken by the authors show that at low feed rates the powder in zone 1 is in a state of continuous agitation. The degree of agitation of a given volume of powder in zone 1 may be expected to increase as the roll speed increases, i.e., as the velocity of the escaping air increases and, at a fixed speed, to decrease as the volume of powder is increased. Worn [10] has shown that an increase in air pressure just above the powder mass may be detected with a manometer; he also draws attention to the fact that the agitation effect is reduced if the air is replaced by hydrogen, which has a lower viscosity.

Furthermore, irrespective of the feed rate, the time-under-maximum-load decreases as the roll speed increases, so that the density of strip might be expected to decrease in a manner similar to that shown to occur for die-pressed compacts.

Naeser and Zirm have reported [11], and the present authors have confirmed, that the density and thickness of strip may be increased by increasing the roll diameter. This method does not depend on increasing the roll/powder friction.

The thickness and density of strip rolled from powder is clearly influenced not only by the roll/powder friction, but also by the behavior of the powder in motion; consequently, the flow of metal powders has been studied and the results are presented below.

VI. Flow and Compaction Properties

The factors that prompted a study of the relation of the flow of metal powders to their behavior in compaction by rolling were: (a) the discrepancy between the maximum gripping angle obtained for copper powder and the gripping angle for a solid copper bar;

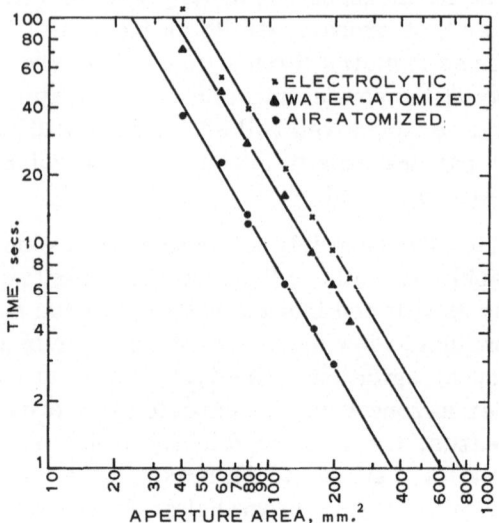

Fig. 11. Relationship between time of flow and area
of aperture when M = 800 g and θ = 60°.

(b) the variation in gripping angle with the powder particle shape;
(c) the fact that for a given particle shape the gripping angle appears to be independent of the mean particle size, whereas the thickness of the strip produced decreases with the mean particle size; (d) the fact that very fine powders can only be compacted at low rolling speeds; and, (e) the fact that spherical particles could not be compacted at all.

A measure was sought of an attribute of a mass of particles in relative motion, thought to be dependent on time and on interparticle contact area, i.e., analogous to viscosity rather than to friction. This was achieved by measuring powder flow.

The apparatus consists of a brass hopper with vertical sides and an inclined floor, giving access to an aperture of fixed width but variable depth whose area can be accurately determined. The whole is supported on an adjustable tripod. The hopper was loaded in a standardized manner with as large a volume of powder as could be accommodated. The time for all the powder to pass the aperture was measured to ± 0.1 sec, and the mean of at least three such measurements was taken as being the flow time.

Experiments made with electrolytic, water-atomized, and air-atomized powders (E0, W0, A0, respectively), showed that the time of flow t is related to the initial weight of powder M, the area of the aperture A, and the inclination of the bottom of the hopper to the horizontal θ, by an equation of the form

$$t = \frac{kM}{A^n \sin \theta} \tag{3}$$

where k varies with the type of powder and n would appear to be constant within the limits of experimental error (Fig. 11). Apart from the $\sin \theta$ term, this relationship is similar to that found by Leadbeater and his co-workers [12] for the flow of iron powders. The $\sin \theta$ term was not included in their results, because the hoppers they used took the form of inverted truncated cones of standard angle θ.

The values of k for the powders E0, W0, and A0 obtained from equation (3) when A = 100 mm^2, n = 1.64 (mean value for powders E0, W0, A0), $\sin \theta$ = 0.866, and M = 800 g, are given in Table II. The values of k for E0, W0, and A0 are in the approximate ratio 3 : 2 : 1, respectively.

The term k in the equation of flow is the "viscosity" analog sought in these experiments. The term "powder-viscosity factor" will be used to describe k, since the term "flow factor" used by Leadbeater and his co-workers may be confused with the test in common industrial use that determines the time taken for a given weight of powder to flow through an inverted truncated cone of known dimensions. The analogy seems justifiable, since the communication or impedance of motion between particles is more akin to viscosity than to friction. A fast-moving particle will communicate some of its momentum, linear or angular or both, to an adjacent particle of lower momentum. To describe the effect in terms of friction alone would not imply a transfer of momentum.

In the equation for powder flow, k and n are constant for a given powder, and the effect of particle shape on k has already been tabulated (Table II). Other variables associated with metal powders that may affect the flow properties are particle size, size distribution, and the surface condition. The last would probably be influenced by the presence or absence of an oxide film or adsorbed films of any kind, including moisture, the amount of which

Table II. Flow Time, t,
and Powder Viscosity Factor, k

Powder	$E0$	$W0$	$A0$
t, sec.	27·8	19·0	8·8
k	57·9	39·6	18·3

Table III. Powder-Viscosity Factor, (k)
and Strip Thickness

Powder Designation	Flow Time (t), sec.*	Mean Value of n	k	Average Strip Thickness, mm. (Initial roll gap $= - 0 \cdot 1$ mm.)
$E1$	22·0 ⎫		34·4 ⎧	0·72
$E2$	25·2 ⎬	1·58	39·5 ⎨	0·62
$E3$	42·0 ⎭		65·8 ⎩	0·38
$W1$	19·5 ⎫		30·5 ⎧	0·55
$W2$	21·0 ⎬	1·58	32·9 ⎨	0·65
$W3$	26·7 ⎭		41·8 ⎩	0·45
$A1$ ⎫ (air-	9·0 ⎫		14·1 ⎧	...
$A2$ ⎬ atomized	9·8 ⎬	1·58	15·4 ⎨	...
$A3$ ⎭ powder)	13·9 ⎭		21·7 ⎩	...

*Flow time when M = 800 g, θ = 60°, aperture = 100 mm^2.

would vary with the humidity of the atmosphere. Further experiments have shown that for particles of a given general shape, k increases as the average particle size decreases, the increase being more pronounced at average particle sizes of less than 100 μ.

The relationship between k and the thickness of strip rolled under fixed conditions from the electrolytic and water-atomized powders of Table III (see also Table II, p. 166) is shown in Fig. 12. There is a fairly good correlation between the two for the electrolytic powders, but not for the water-atomized powders, one of which (W2) had become fairly heavily oxidized. The anomalous increase in thickness shown by the oxidized powder may be due to increased roll/powder friction similar to that occurring with some oxidized solid metals. No corresponding increase in the gripping angle for this powder could be detected; it was probably below the limits of error.

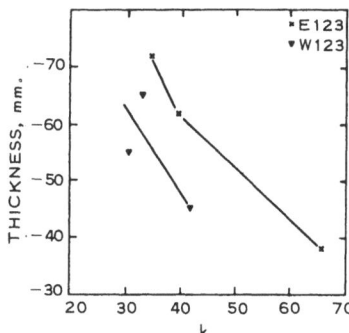

Fig. 12. Relationship between strip thickness and powder-viscosity factor (k) for electrolytic (E 1,2,3) and water-atomized (W 1,2,3) powders.

Although k would not appear to indicate the thickness of strip that can be rolled from a given powder irrespective of the surface condition of the latter, it may serve as a "criterion of cohesion" with respect to given rolling conditions. The following example serves to clarify this point. Two copper powders were obtained which could only be compacted by the most rigorous control of the roll speed and roll gap. One was a mixture of electrolytic and air-atomized powders with good flow properties, i.e., low k; the other was a very fine powder produced by the Chemetals process, with rather poor flow properties, i.e., high k. The first produced a weak green strip through lack of sufficient interparticle bonds; the second, a weak strip because its flow rate was too low to keep pace with the minimum roll speed that could be used, so that inadequate feed occurred. The two limits of k were thus defined within which a powder could be expected to yield a coherent strip under the rolling conditions used in these experiments. It was found that the k values for other copper powders which had been successfully compacted lay within these limits. Furthermore, it can be seen that even those powders whose shape and size distribution will yield a strong green compact under conditions of static pressing must possess flow properties that will enable the powder feed to keep pace with the roll speed.

VII.   The Mechanism of Compaction

Any elaboration of the mechanism of compaction must explain the following phenomena:

(1) A given powder is not compacted if the roll gap is too large.

(2) Increasing the powder feed increases the thickness and density of the strip produced at a given roll gap.

(3) Increasing the roll speed decreases the density and thickness of strip produced at a given roll gap.

(4) The green strip shows better bonding across vertical planes parallel to the rolling direction than across vertical planes parallel to the transverse direction.

(5) The gripping angle varies for particles of different general shape.

(6) The gripping angle is approximately constant for powders of a given shape but varying average particle size.

From the experimental evidence, the following explanations may be advanced:

(1) The powder must undergo a minimum degree of compaction if a coherent strip is to be produced. For a given powder the compaction ratio is determined by the gripping angle and the roll gap and, since the gripping angle is fixed, increasing the roll gap decreases the compaction ratio until eventually a point is reached at which the powder is no longer compacted.

(2) Increasing the powder feed increases the packing density of the powder in the roll gap by reducing the agitation it experiences, so that for a given compaction ratio the density of the strip is increased. As the density is increased the force separating the rolls must increase, resulting in roll flattening and a thicker strip.

(3) Increasing the roll speed reduces the time during which the powder is subjected to a compressive load, thus decreasing the density of the strip. Furthermore, as the roll speed increases the velocity of escape of entrapped air must increase and the superincumbent weight of powder must decrease, so that the powder density is decreased. This, too, reduces the density of the strip.

(4) It is clear from the results of the electrical resistance measurements in the two principal directions (see p. 147) that the particles or particle surfaces undergo relative movement in the rolling direction. Relative movement of the particle surfaces alone could occur only when the particles were already intimately packed and undergoing deformation in the rolling direction. On the other hand,

relative movement of the discrete particles might take the form of a rotary motion, induced by the couple consisting of the friction between the rolls and the adjacent layer of particles and the constriction of the roll gap. Relative motion would then be greater between horizontally adjacent particles than between vertically adjacent particles, which would tend to "gear" together, and there would then be a more vigorous abrasive effect in planes parallel to the rolling direction than in the other two principal planes. This would tend to remove surface films and so permit improved bonding to occur across the one plane.

(5) The rotary motion would be communicated through the powder by interparticle friction, and slipping between the particles would give rise to a powder-viscosity effect, whereby the force along surfaces concentric with the roll surfaces would decrease as the distance from the latter increased. With increasingly irregular powder surfaces the effect would persist for an increasing distance from each roll surface, and the gripping angle would be determined by the point in the roll gap where the rotary motion imparted to the powder by one roll surface met that imparted to the powder by the other roll surface. Thus, the gripping angle would increase with increasingly irregular particle surfaces.

(6) To a first approximation the gripping angle is determined by the roll/powder friction, which in turn has been shown to be dependent on powder-particle shape. The only variable associated with particle size is the powder-viscosity factor, and it would appear that, in the absence of variations in particle-surface conditions capable of affecting the roll/powder friction, the powder-viscosity factor determines the strip thickness — mainly by its control of the internal powder-feed rate.

Acknowledgment

This work was carried out at The Goldsmiths' Laboratory of the Department of Metallurgy in the University of Cambridge. The authors are grateful to Professor G. Wesley Austin, at whose instigation the work was undertaken, for his continued interest and encouragement.

References

1. E. Orowan, Proc. Inst. Mech. Eng., 150 :140 (1943).
2. C. L. Smith, F. H. Scott, and W. Sylwestrowicz, J. Iron Steel Inst., 170 : 347 (1952).
3. G. Wallquist, ibid., 177 :142 (1954).
4. M. Cook and R. J. Parker, J. Inst. Metals, 82 :129 (1953-1954).
5. R. B. Sims, J. Iron Steel Inst., 178 : 19 (1954).
6. L. R. Underwood, The Rolling of Metals, Vol. I, London, Chapman and Hall (1950), pp. 10 et seq.
7. L. R. Underwood, op. cit., p. 146 [after W. Lueg, Stahl Eisen, 55 :1105 (1935)].
8. L. R. Underwood, op. cit., p. 13.
9. L. R. Underwood, op. cit., p. 138 [after W. Tafel and E. Schneider, Stahl Eisen, 44 : 305 (1924)].
10. D. K. Worn, Powder Met., No. 1/2 :85 (1958).
11. G. Naeser and F. Zirm, Stahl Eisen, 70 : 995 (1950).
12. C. J. Leadbeater, L. Northcott, and F. Hargreaves, Powder Metallurgy (Selected Government Research Reports, Vol. 9), London, H. M. Stationery Office (1951).

Chapter 9. Extrusion of Powders I

The Extrusion of Metal Powders

Norman R. Gardner, Allan D. Donaldson,
and Frank M. Yans

Metalonics Corporation
a Subsidiary of Kawecki Chemical Co.

Introduction

The demands of an advancing technology are providing an opportunity for metal powder producers and organizations which are competent in extrusion to jointly supply products to meet its requirements. The technical and perhaps economic attributes of using particulate material will best be achieved when methods of fabrication are available to consolidate the materials into useful shapes at normal metal-working costs. Extrusion is perhaps ideally suited to meet this need in many cases, particularly when the length to diameter ratio is high as in rods, tubes, and special shapes. In extrusion, we find a process suited to prototype, small volume production in the 100-lb class as well as tonnage output. Extrusion is also an economically attractive process, insofar as it offers simultaneous densification and reduction in area of from 6 to over 100 times in a single operation. No other process can yield deformation of this magnitude and under conditions so ideally suited to powders, i.e., high compressive forces. These high unit pressures (up to 250,000 psi), which are characteristic of extrusion, enable powders to be densified and deformed at temperatures which are lower than those required for sintering or even hot pressing. The high reduction ratios result in a large increase in interfacial particle area, as well as exposing relatively active oxide free surfaces. Both of these effects assist in diffusion to form homogeneous alloys, where such are desired.

One of the primary technical advantages of utilizing powder materials is the flexibility arising from the lack of phase diagram and melting point restrictions or consequences, resulting in materials with unique mechanical or metallurgical properties. The possibilities of pre-engineering materials in billet form and reducing the billets to some configuration close to the desired final size is the major contribution offered by the extrusion process.

Extrusion Process

The desired powder product is, mechanically speaking, subjected to the following sequence of events, as shown in part in Fig. 1.

1. Blending in the case of alloys, or dispensions.
2. Prior compaction.
3. Canning.
4. Evacuation and seal-off.
5. Heating and lubrication.
6. Pressure application
   a. deformation to fill voids in powder or billet clearances.
   b. bulk density and hydrostatic forces generated.
   c. metal deformation through die aperture and streamline flow.
7. Dejacketing.
8. Sizing or secondary working.

Perhaps the most singularly attractive aspect of the extrusion process as applied to metal powders is the fact that 100% densities can be achieved almost invariably. This can be accomplished without the use of any sintering procedures whatsoever.

The grain size and shape of an extruded product made from powders is usually quite characteristic, i.e., quite small and elongated. For example, if a billet of 200 mesh powder (74-μ par-

ticle) is extruded at a temperature at which extensive grain growth
is not allowed to occur and a reduction ratio of 50 : 1 the extruded
product will have a grain 10 μ in diameter and about 300 μ long.
The reason the grains are not longer is that a slight degree of recrystallization usually occurs which subdivides the original powder particle along its length. It should be noted that the extrusion
process has increased the 200 mesh particles surface area by a
factor of 7.

This particular decrease in size and increase in surface area
allows interparticle diffusion to occur at a drastically increased
rate (50 to 100 times faster) and also exposes new, oxide-free
surfaces. This more intimate interaction between particles usually promotes increased strength and formability in the final product.

Where previously compacted and completely densified starting materials are employed, the extrusion of powder metal billets
does not differ substantially from the extrusion of their cast
counterparts. However, when such compaction and densification
are accomplished by conventional pressing and/or sintering techniques, attention must be paid to whether full density has been
achieved in order to accurately calculate the physical reductions
required and to ensure that internal oxidation of the billet at temperature does not occur.

It should be borne in mind that extrusion or vertical forging
presses are particularly well suited to convert powders to billet
stock either during the actual extrusion operation or as a separate
step. In many cases, insofar as extrusion presses are generally
disposed horizontally, forging presses are more useful in hot
pressing of metal powders. Extrusion presses range up to 12,000
tons capacity and the forging presses up to 50,000 tons. Such
equipment would be readily available for tonnage production without investment on the part of powder producer or consumer, and
we expect would be competitive with multiple pressing and sintering procedures. The availability of higher capacity hydrostatic
pressing units will also make extrusion billets more readily available as well as increase the maximum sizes obtainable.

Economically, the most attractive process would involve
simply pouring loose powders into the extrusion container and

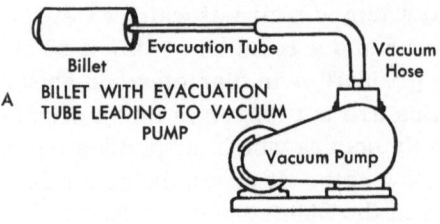

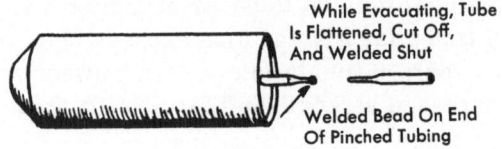

Fig. 1A and B. Evacuation and sealing of powder extrusion billet.

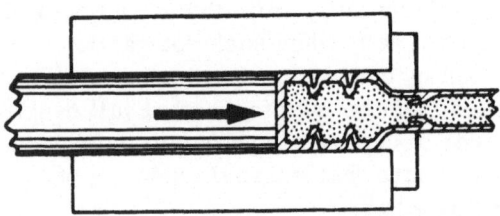

Fig. 1C.   Folding of metal can with loosely packed powders.

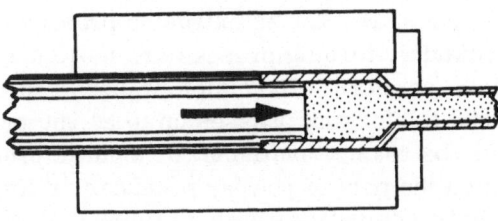

Fig. 1D.   Penetrator technique to avoid folding of can.

pressing these powders directly through the die. However, a number of facts militate against this utopia. First, most metals must be heated in order to extrude them and this would indeed be a problem with loose powders, especially if oxidation rates were rapid. Powders could be heated in the extrusion container, but this is quite slow and very costly, since expensive presses are tied up as well as being limited to extrusion tooling tempering temperatures (900°F). In addition, lubricating loose or uncontained powder gives rise to special problems including the infiltration of the lubricants into the powder billet. Extrusion of magnesium powders and certain solders by this procedure is, however, reported to be practical.

We believe that for the near future something in between direct extrusion of particulate materials and bare sintered billets will be worth investigating. Namely, the extrusion of canned powders which have some degree of prior cold or hot compaction. Two reasons for this are that most materials of technical interest require extrusion temperatures in excess of 900°F and that techniques have been reasonably well developed to simultaneously compact and extrude powders in the extrusion operation, thereby avoiding the necessity and associated costs of pressing and sintering. (See Fig. 1.)   The can usually serves numerous functions concurrently. For instance, it may be used as a processing vessel for evacuation, dehydriding, or annealing prior to extrusion. Further, the canning insures that the materials can be processed without contamination from atmospheric reactions and/or lubricants. The sheath serves the added functions of facilitating the flow of metal through the extrusion die and, in the case of reactive materials, prevents seizing with the tooling and contamination from the atmosphere as the shape exits from the die.   The choice of a particular cladding material is dictated by (1) its hot plasticity relative to the material to be extruded; (2) the possibilities for objectionable alloying or the formation of low-melting phases, (3) the ease with which it may be removed by chemical or physical means, and, (4) material costs. For increasingly high extrusion temperatures, magnesium, aluminum, copper, steel, and molybdenum are considered in that order.

Independent of the can material, difficulties can be anticipated in extruding canned powder billets in which the powder core has a very low density. It is necessary to insure that the deformation of the billet does not result in folding in of the cladding, as shown in Fig. 1. This may be accomplished by preferentially deforming or densifying the powder core prior to exerting pressure on the can. Such a sequence is achieved by using a solid penetrator or a friable "dummy block." The penetrator is designed so as to traverse through the can to a point where a density approaching theoretical is realized. At this stage, pressure is exerted almost evenly on the core and can.

The use of a metal sheath facilitates the extrusion of relatively brittle materials, as discussed in more detail later, by absorbing the major amount of tensile forces that exist at the clad-die interface as the extrusion exits from the press. Tendencies for surface cracks to develop are minimized and deformation of the core is accomplished by relatively pure compression. The cladding material is generally less resistant to deformation during extrusion than is the powder core.

The force required in extrusion is given by $F = AK \ln R$ (where F is expressed in tons; A is the area of the container in square inches; K is the extrusion constant, or a measure of the stiffness at temperature, in tons per square inch; and R is the extrusion reduction or the ratio of the initial area of the billet relative to the cross-sectional area of the extruded shape). For composite or clad billets, the total Force (limited by press capacity) required is given by $F = (A \text{ clad } K \text{ clad} + A \text{ core } K \text{ core}) \ln R$. Materials which are not amenable to high amounts of cold work after extrusion may often be extruded close to final size by use of a thick can, commonly known as the filled-billet technique. Also, conversions can be carried out on lower capacity presses which are usually less expensive to operate. In general, the sheath material is removed chemically in suitable hot acid solutions. However, it is sometimes desirable to leave the sheath intact so as to have a composite rod or special shape, the surface of which may have certain desirable characteristics relative to the core or powder product.

Figure 2 provides a rough indication of the relative stiffness or resistance to plastic deformation of various metals as a function of extrusion temperature.

Products – Pure Metals

The extrusion of pure metal powders is of interest for one or more of the following reasons: (1) if the resultant product has properties which are superior to extrusions made from castings, (2) if the metal cannot be easily cast into billet form; or (3) if there is an economic advantage. The latter case rests largely on the cost of producing powder relative to the cost of producing ingot. Until recently, castings have had a clear advantage over powder, in most cases. However, there appears to be a trend toward hydrometallurgical and gas-reduction processes which result in powder of high quality at competitive prices. The copper operation in the Philippines bears watching, as does the direct reduction of iron and nickel.

A metal such as tungsten, owing to its exceedingly high melting point (6170°F) has been processed almost exclusively by powder metallurgy means. The casting of tungsten, while now feasible, leads to no overall advantage, and recent developments have indicated that tungsten powders can be hot-extruded to yield rod and tubing.

Powder processing of molybdenum (melting point 4750°F) remains of commercial importance despite efficient vacuum melting procedures and a trend toward castable alloys. The large grain size of the cast product, however, results in poor mechanical properties. Accordingly, when cast molybdenum is considered, breakdown operations such as forging or pre-extrusion are mandatory to reduce the ultimate grain size and provide adequate mechanical properties. Starting with a sintered powder product affords an opportunity to eliminate many primary working operations and to arrive at a final product more directly.

Processing of beryllium to yield rod, tubing, and shapes of sufficient ductility to be at all useful in structural-type applications dictates the use of powder starting material. The powder, generally minus 200 mesh, insures that the grain size to begin with is of proper size and the thin oxide layer (about 1% BeO on each particle) provides a barrier for grain growth during hot working. Figure 3 indicates the grain size of beryllium extruded from cast and powder billets, respectively. The large reductions that can be achieved by means of extrusion which, in the case of beryl-

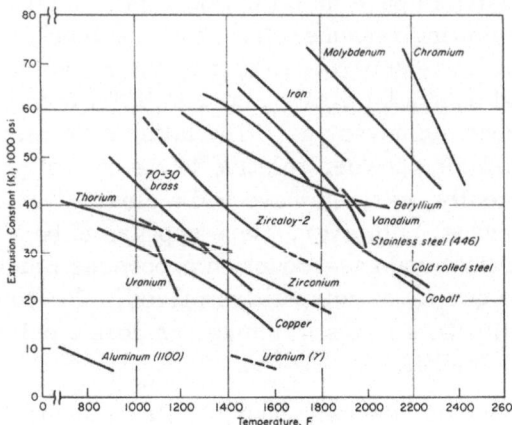

Fig. 2. Extrusion constant versus temperature.

Fig. 3. Microstructure of beryllium metal extruded from a casting (left) and powder (right). Transverse section.

lium may be up to 50 times reduction in area, makes it feasible in a single operation to introduce a high degree of preferred orientation. During extrusion the hexagonal beryllium crystal orients itself in such a manner that its "C" axis is perpendicular to the extrusion direction and above a reduction of approximately 6 times, elongations of approximately 20% are obtained in the direction of extrusion. For a material such as beryllium, where conventional working at room temperature is not possible, extrusion is an ideal process and its utilization is increasing rapidly.

Dispersion-Hardened Metals and Alloys

A voluminous amount of data is available to clearly indicate the excellent strength and stress rupture properties of dispersion-strengthened metals and alloys. In general, it is probably safe to say that the major problem is now to fabricate these materials into useful shapes and to broaden their range of application and their attributes. The same characteristics which in part lead to the interest in these materials makes for problems in working them by conventional means. High yield strengths, low ductility, high annealing temperatures, and high hardness and abrasiveness all lead to fabricating problems. Accordingly, it is desirable to hot-work the material to as close to final size as it possible. The direct rolling of strip and sheet and extrusion of rod, tubing, and shapes provide a means of accomplishing this objective.

In addition to problems associated with fabrication, increased attention to providing more well-balanced materials is needed. For instance, it is not adequate to have a material with superb high-temperature properties only to find that at those temperatures it cannot withstand other environmental conditions such as oxidation, thermal shock, and cycling. Further, wider application could be expected if the properties in a lower temperature range were also superior. There is a natural trend therefore, to superimpose onto the high-temperature characteristics of dispersion-hardened metals the previously useful attributes at the lower temperature range arising from solid solution and precipitation hardening as well as cold work. Greater attention to cold work is suggested insofar as it seems clear that through dispersion hardening, the effects of cold work are retained due to a substantial increase in the recrystallization temperature.

The interest in alloy powders and their increasing availability will assist in broadening the useful temperature range, particularly at lower temperatures. We believe that the use of elemental powders to form alloys may be of special interest in facilitating the fabrication of dispersion-hardened alloys. For instance, it may be reasonable to expect that if complete alloying does not occur or can be kept from occurring by employing low extrusion temperatures, both extrusion and further fabrication will be aided.

The extrusion of dispersion-hardened powders is facilitated by the use of a metal can. For instance, it is reported* that aluminum—aluminum oxide extrusions at a reduction ratio of 15 times and a temperature of 500-600°F required an extrusion pressure of 70 tons per square inch. The use of a pure aluminum can under similar circumstances resulted† in a reduction in extrusion pressure to 35 tons per square inch due to improved lubrication. In addition to utilizing the can to effectively reduce the friction during extrusion, the sheath may purposely be designed and introduced into the process to solve certain other problems. Typically, where the core of dispersion-hardened material is designed primarily to afford high-temperature mechanical properties. This cladding may be chosen for its special surface characteristics such as oxidation resistance. The ability of the extrusion process to introduce such a surface concurrent with the fabrication of the dispersion-hardened core, i.e., via coextrusion, has not been sufficiently explored.

Alloys

Super Alloys

A potentially fertile area for investigation involves the manufacture of super alloys, including such nickel-base materials as Udimet 500, Rene 41, Inconel, and the Nimonic series. These materials are all characterized by a hardening mechanism involving aluminum—titanium compounds which precipitate during an aging process. In general, they are made by organizations which are oriented toward the casting process. Accordingly, there is

---

*A. Von Zeerleder, Plansee Proc. (1952), p. 211.
† British Patent No. 734,778.

little incentive on their part to investigate the powder route. A number of problems are intrinsically associated with casting these alloys. First, a serious problem exists in undesirable segregation during solidification of some of the components. Next, the alloys once cast are difficult to further fabricate. The Rene 41 alloy, in particular, is only capable of safely absorbing about 15% cold work between anneals. During hot working, as by rolling, the alloy must be fabricated in a relatively narrow temperature range, i. e., 1850-2150°F. Such a temperature range is difficult to maintain in practice. When a temperature is experienced in excess of 2150°F, the alloy is hot short.

We have found that producing Rene 41 by means of blending and extruding elemental powders at a temperature of 1700°F, and a reduction of 16 : 1, we can cold-work the resulting product over 70%. This radical improvement in workability both at elevated and room temperature is of considerable significance where wire, tubing, or special high tolerance shapes are required. In addition to the possibility of producing the present alloys via the powder route, a greater flexibility exists which can permit the manufacture of more highly alloyed materials with even better properties. It should not be difficult, for instance, to increase the aluminum and titanium levels above the presently limited $1^1/_2$% and 3%, respectively. After final fabrication, the materials can be diffusion heat-treated to yield the desired properties.

Hard-Facing Alloys

At the present time many of these materials are produced by casting directly to final sizes ($^1/_8$-$^1/_4$-in.-diameter rod). Such finger castings in the cobalt and nickel base alloys are limited in their length-to-diameter ratio. In addition, segregation and voids in the casting are prevalent and give rise to problems during welding. Attempts are being made to determine the feasibility of producing some of these materials directly from powders blended in the proper alloy proportions. In this type of application it is not clear that complete alloying is essential to the final product insofar as the material is made molten during the welding operation. Here, again, the possibility of producing improved materials which are not limited by the various restrictions associated with a melting operation is worthy of attention.

Time does not permit a more detailed discussion of the process or its many other applications, such as a host of welding rod compositions, cermets characterized by small metal additions, and so forth. It appears clear that in order to meet new materials requirements and to facilitate the production of metal and alloy shapes which are difficult to produce by casting technology, extrusion provides an ideal means for powder processing and fabrication. It is important to bring the added dimension of metalworking technology to bear on these problems. The unique technical and economic advantages of the extrusion operation are just beginning to be utilized, and it is expected that increased expenditure of imagination and effort will soon bear fruit.

Chapter 10. Extrusion of Powders II

Hot-Extruded Chromium Composite Powder

R. V. Watkins, G. C. Reed, and W. L. Schalliol*

Bendix Products
Aerospace Division
South Bend, Indiana

Introduction

A chromium—magnesium oxide composite was developed in 1960 by The Bendix Corporation for use in high-velocity oxidizing environments at temperatures between 1100-1700°C. This composite system exhibits room-temperature ductility even in the re-crystallized condition. The normal rod-processing procedures for these composites have been previously reported [1]. They involve sintering pressed powders and hot extrusion at 1100-1300°C of the nickel-clad billets. Hot extrusion of the pressed composite powders followed by sintering was considered as an alternate processing technique for producing rod. An investigation with this approach is the subject of this paper.

---

* This work was done by the authors at Bendix Products Aerospace Division, South Bend, Indiana. Dr. Schalliol, who was previously Director of Metal-Ceramic Research at Bendix, is now with CTS Corporation of Elkhart, Indiana.

Fabrication of Rod

The room compositions shown in Table I were fabricated into rod using generally the method for chromium powders reported by Loewenstein et al. ([2], p. 570). A commercial electrolytic chromium was obtained as −325 mesh material with a Fisher sub-sieve size of 14.4 μ.

All powders were mechanically blended and then sized through a 200 mesh screen. The −200 mesh powders were precompacted to 25,000 lb/in.^{2}, screened through a 30 mesh screen, vibrated into a rubber casing, and evacuated before the hydrostatic pressing operation. The powders were hydrostatically pressed at 20,000 lb/in.^{2}, and the compacts were outgassed in a vacuum furnace at 540°C for 1 hr. No water or binder is needed in this series of operations. The rough compact was then machined to fit inside a stainless steel container. Each container consisted of four parts: a shaped nose, a stainless steel tube, a lid with a small hole in it, and a capillary tube. After the nose was welded to the tube, the machined compact was fitted into it and the end cap with the small degassing tube attached was welded in place. A small depression machined in the end of the billet allowed out-gassing without danger of halting the flow of air. This was accomplished by evacuation on a manifold to a pressure of 15 μ. Heating the whole assembly to 320°C facilitated this step. This pressure and temperature were maintained for at least 48 hr to insure thorough evacuation. At the end of this time the tube was welded shut and removed from the manifold.

Billets 3 in. in diameter and 6 in. long were heated to 1200°C, held for 1 hr, and extruded by the Metals and Ceramics Laboratory of the Aeronautical Systems Division, Wright-Patterson AFB, Ohio. An extrusion ratio of 10:1 was used successfully on all extrusions. Corning 0010 borosilicate glass was the lubricant. The extrusion pressures in pounds per square inch are shown in Table II. The extrusions were approximately 3 ft long and 1 in. diameter, and all were free of visible defects.

Sintering

The effect of sintering subsequent to extrusion was studied after heating samples of rod in vacuum or hydrogen. Vacuum heat

Table I. Nominal Composition of Hot Extruded Powders*

Extrusion Numbers	Composition	Cr	MgO	Ti	Nb
EA100 and EA101	Electrolytic Cr	100.0			
EA102 and EA103	Cr-MgO Composite	93.5	6.0	0.5	
EA104 and EA105	Cr-Nb-MgO Composite	93.0	6.0		1.0

*Powder Used: Cr — Union Carbide, −325 Mesh Elchrome Grade, Lot 37154
MgO — Morton Chemical, −325 Mesh, U.S.P. Light Grade
Ti — Charles Hardy, −325 Mesh, Electrolytic Grade B
Nb — Union Carbide, −325 Mesh, Columbium Metal Grade

Table II. Extrusion Pressures of the
Chromium Composites Sintering

Extrusion Number	Composition	Pressure (PSI)
EA100	Electrolytic Cr	106,000
EA101	Electrolytic Cr	107,000
EA102	Cr-MgO Composite	123,000
EA103	Cr-MgO Composite	118,000
EA104	Cr-Nb-MgO Composite	108,000
EA105	Cr-Nb-MgO Composite	103,000

treatments were performed at 1×10^{-3} torr at maximum temperature. The atmosphere used in the hydrogen treatments had a dew
point of −50°C after passing through a palladium diffusion purifier.
Results of chemical analyses of rod samples after extrusion and
sintering are given in Table III. The microstructure changes for
electrolytic chromium and the Cr−MgO composite sintered in
vacuum are shown in Fig. 1.

The as-extruded structure remained essentially unchanged;
even after sintering up through 980°C. The average grain size of
the chromium matrix in Figs. 1a and 1c is about 7μ. Higher sintering temperatures resulted in coarsening of the grains and ag-
glomeration of the nonmetallics at the boundaries. The MgO additive in the composite restricted grain growth of the chromium
matrix. This can be seen in Fig. 1b and 1d, which show the typical structure of samples sintered at 1540°C.

Sintering in hydrogen was done to reduce oxides present and
to avoid material loss, because of the high vapor pressure of the
chromium. The hydrogen sintering achieved both results and at
1540°C produced a microstructure in electrolytic chromium con-

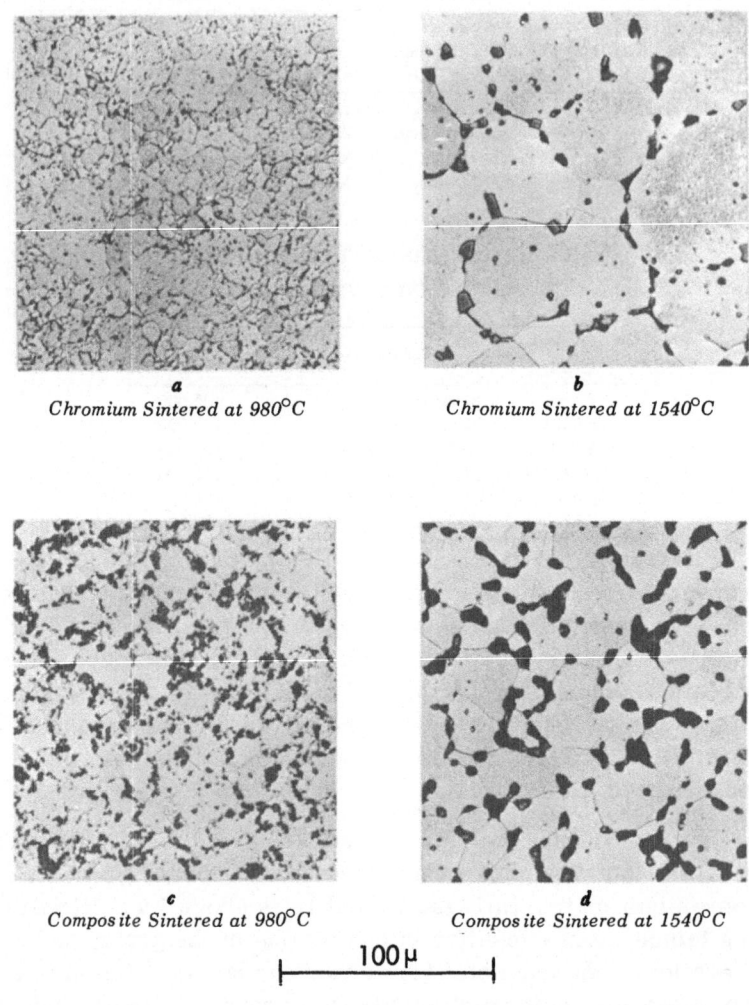

a
Chromium Sintered at 980°C

b
Chromium Sintered at 1540°C

c
Composite Sintered at 980°C

d
Composite Sintered at 1540°C

100 μ

Fig. 1. Microstructures of pure chromium (a and b) and chromium—magnesium oxide composite (c and d) after sintering for 2 hr in vacuum (original magnification 500 ×). Etchant: 3 parts glycerin, 1 part HCl.

Table III. Results of Chemical Analyses

Sample	O₂ %	N₂ PPM	C PPM	Fe PPM	S PPM	Mg %	Ti %	Nb %
Starting Cr Powder	.67	52	202	130	150			
As-Extruded Cr	.60	44	238					
Cr Sintered 2 Hours 1600°C Vacuum	.76	64	99					
Cr Sintered 2 Hours 1600°C Hydrogen	.31	77	47					
As-Extruded Cr-MgO	2.73	27	300			3.47	.41	
As-Extruded Cr-Nb-MgO	2.65	69	456			3.50		.99

Table IV. Effect of Sintering Temperature on Density*

| Material............ | Electrolytic Chromium (Density~g/cc) | | Cr-MgO Composite (Density~g/cc) | | Cr-Nb-MgO Composite (Density~g/cc) | |
Sintering Atmosphere for 2 Hrs... Sintering Temperature °C	Hydrogen	Vacuum	Hydrogen	Vacuum	Hydrogen	Vacuum
As Extruded	7.10	7.10	6.45	6.45	6.50	6.50
980	7.09	7.11	6.49	6.54	6.51	6.47
1320	7.10	7.00	6.50	6.43	6.20	6.39
1430	6.97	6.89	6.37	6.37	6.20	6.18
1540	6.69	6.72	6.23	6.18	5.98	5.66
1600	6.55	6.55	5.86	6.07	5.85	5.81
Average % Change From As-Extruded Density	7.8		7.6		10.3	

*Archimedes' Method

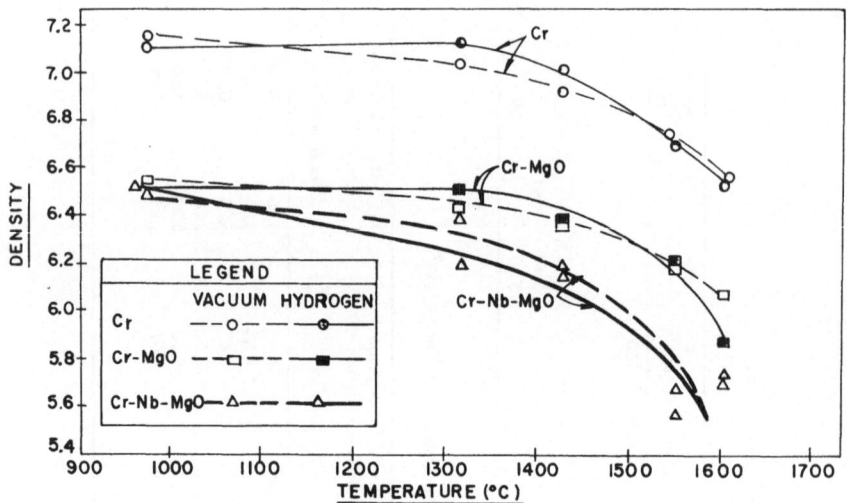

Fig. 2. Effect of sintering temperature on density.

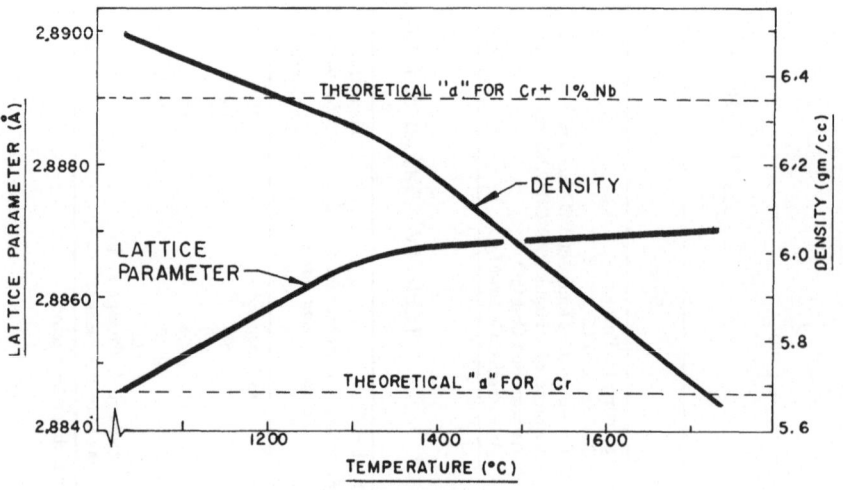

Fig. 3. Effect of temperature on lattice expansion and density of Cr—Nb—MgO in hydrogen.

taining many voids where oxides had been previously located. The reduction in oxygen and carbon content is shown in Table III for the hydrogen-sintered chromium.

Density determinations were made on samples of each rod after post-extrusion sintering for 2 hr in either vacuum or hydrogen. The results are given in Table IV and shown graphically in Fig. 2. Lattice parameter measurements of the Cr−Nb−MgO composite were made after sintering samples in hydrogen at temperatures up to 1680°C. Figure 3 shows the lattice expansion which occurred and compares it with the density change for the same samples. Figure 4 shows the progressive disappearance of a particle of Nb as the Cr−Nb−MgO composite was heated in vacuum. The voids result from the rapid diffusion of Nb into the chromium matrix and help to explain the decrease in density which occurs during sintering.

Tensile Tests

Previous work on $\frac{5}{8}$-in.-diameter extrusions made at Bendix in the same manner as the 1-in. extrusions used in this study revealed that subsequent sintering was necessary to improve the ductility. Table V compares tensile properties at 66°C for chromium and the Cr−MgO composite after various sintering cycles. Following this preliminary study, 1-in.-diameter extrusions were produced and samples were sintered for 2 hr at 1600°C in both vacuum and hydrogen environments. Tensile properties were then determined for specimens machined from these bars. The machinability was "poor" for the electrolytic chromium, "excellent" for the Cr−MgO composite, and "good" for the Cr−MgO−Nb composite. The button-head round tensile bar preparation and testing procedures in air have been reported previously ([2], p. 11]. Table VI shows the effect of test temperature on the tensile properties of chromium-based composites. Since the bulk of the electrolytic chromium rod was used in programs not reported in this paper, elevated temperature tensile properties are not available for this material.

For comparison purposes, reference is made to the work by Metcalfe et al. [3] at Armour Research on extrusion of the same specification chromium powders used for our studies. The fabrication technique was identical except for a lower extrusion tem-

Table V. Effect of Sintering Variables on Tensile Properties at 66°C

	SINTERING VARIABLES			TENSILE PROPERTIES AT 66°C		
Composition	Temp. °C	Time Hrs.	Atmosphere	Yield Strength PSI (0.2% Offset)	Ultimate, PSI	Elongation %
Electrolytic Chromium	1600	2	Vacuum	23,970	29,770	4.1
Electrolytic Chromium	1600	2	Hydrogen	15,670	20,120	3.7
Cr-MgO Composite	980	½	Vacuum	68,250	68,250	2.4
Cr-MgO Composite	1600	½	Vacuum	36,000	50,400	17.6
Cr-MgO Composite	1600	2	Vacuum	29,800	46,100	22.8

Table VI. Effect of Test Temperature on Tensile Properties of Cr-Based Composites

Sintering Cycle.........		2 HRS.—VACUUM—1600°C			2 HRS.—HYDROGEN—1600°C		
Material	Test Temp. °C	0.2% Yield (PSI)	Ultimate (PSI)	Elongation (%)	0.2% Yield (PSI)	Ultimate (PSI)	Elongation (%)
Cr-MgO Composite	27	28,750	47,600	11.9	27,000	40,750	7.6
	66	26,400	44,350	23.1	23,050	42,950	10.9
	150	14,400	31,000	15.6	14,100	29,900	14.6
	500	19,400	33,500	18.9	18,400	34,600	18.2
	1260	3,570	4,140	43.6	2,550	3,280	29.0
Cr-Nb-MgO Composite	27	20,700	23,150	0	22,900	24,500	2.3
	66	20,000	27,750	3.2	18,000	25,450	3.1
	150	14,000	28,650	4.1	17,450	33,050	7.8
	500	18,500	27,330	3.7	21,000	32,600	4.8
	980	26,500	30,080	1.4	25,000	31,450	0.7
	1260	7,700	8,500	7.9	7,750	8,870	3.5

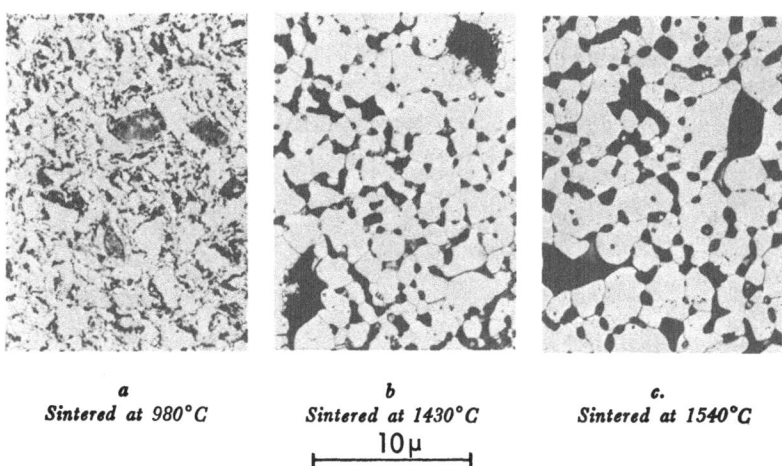

a	*b*	*c.*
Sintered at 980°C	*Sintered at 1430°C*	*Sintered at 1540°C*

$\longleftarrow$ 10 µ $\longrightarrow$

Fig. 4. Progressive disappearance of niobium particles as the Cr—Nb—MgO composite was heated for 2 hr in vacuum (original magnification 500 ×). Etchant: 3 parts glycerin, 1 part HCl.

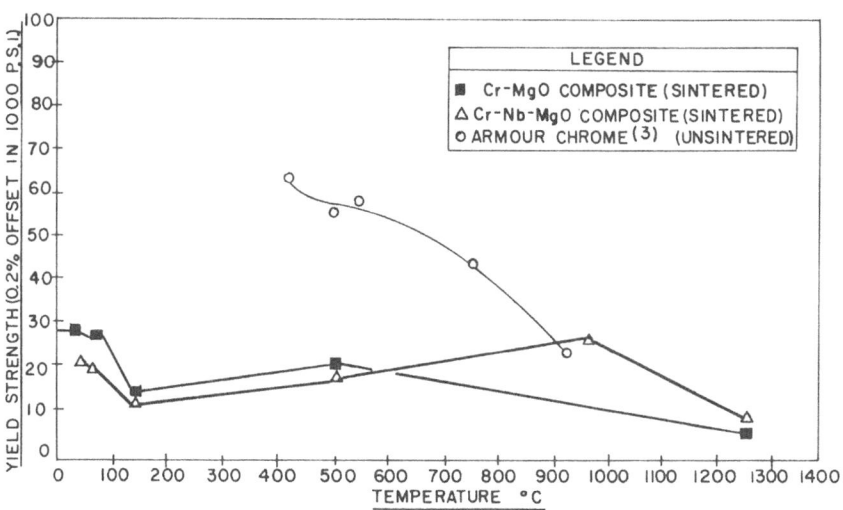

Fig. 5. Yield strength versus test temperature.

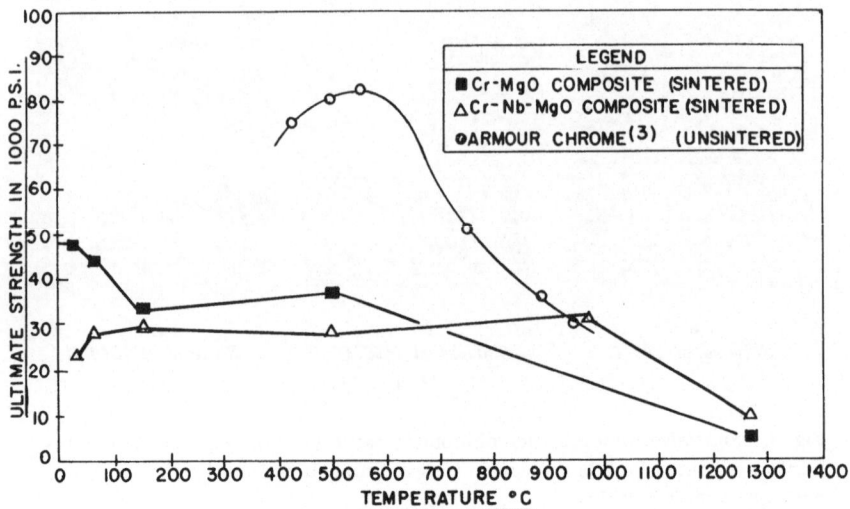

Fig. 6. Ultimate strength versus temperature.

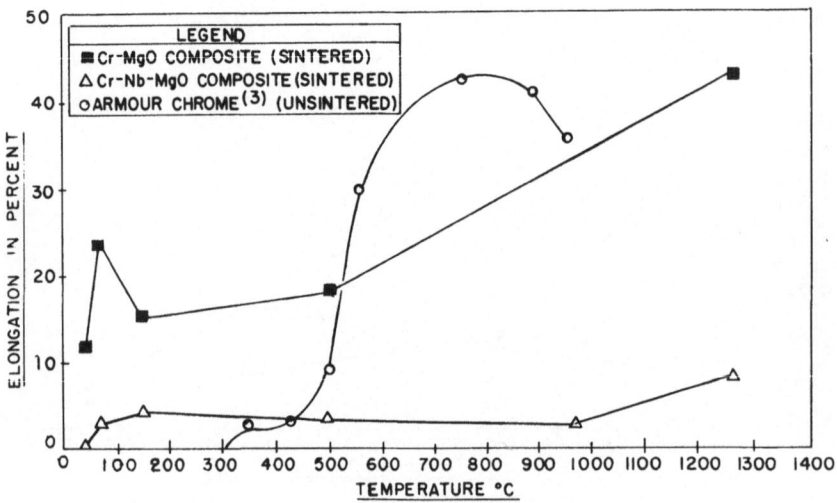

Fig. 7. Elongation versus test temperature.

perature (1000°C) and the use of nitric acid instead of machining
to remove the steel jacket. Figures 5, 6, and 7 compare, respectively, the yield strength, ultimate strength, and elongation of
chromium-based composites with the extruded chromium powder
of Armour ([3], p. 42). The chromium reported by Metcalfe ([2],
p. 8) was extruded below the temperature where recrystallization
and grain growth would occur, which enabled the grain size range
of $1-5\mu$ to be achieved in the as-extruded condition.

Discussion of Results

The addition of MgO to the chromium matrix has produced
the following improvements over electrolytic chromium:

a. Finer-grained microstructure after sintering at 1600°C
    (Fig. 1).
b. Over five times the tensile elongation at 66°C (Table IV).
c. Vastly improved machinability.

The inhibition of grain growth and improved machinability
would have been anticipated by the addition of the insoluble phase
to chromium. The ductility even at room temperature would not
have been anticipated prior to the work reported by Scruggs and
Masterson [4]. A possible mechanism leading to improved ductility is the formation of an impurity sink when the nonmetallics ag-
glomerate during sintering. The $MgO \cdot Cr_2O_3$ spinel has been
identified by x-ray diffraction in the sintered $Cr-MgO$ composite
and this loosely bonded structure could chemically or physically
combine with such troublesome impurities as carbon and nitrogen.
It is hypothesized that this microfluxing cleanses the grain boundaries and agglomerates the impurities into microcells usually at
multigrain interfaces. A comparison between the relative cleanliness of grain boundaries in sintered chromium and the $Cr-MgO$
composite can be made in Figs. 1b and 1d. The reduced nitrogen
content of the as-extruded $Cr-MgO$ composite (Table III) could be
caused by the loss during chemical analysis of a high-nitrogen-
content insoluble complex, but it also supports the hypothesized
cleansing action within the chromium matrix.

The decrease in density with increased sintering temperature
can be attributed to internal gas evolution in the case of chromium

and the Cr—MgO composite and the added Kirkendall effect in the
Cr—Nb—MgO composite. The lattice parameter increase paral-
leling the density decrease shown in Fig. 3 substantiates the latter
reasoning, and the work by Moon and Consolazio [5] on gas evolution explains the former. The porosity observed could be rem-
edied by subsequent hot work.

The tensile properties shown in Figs. 5, 6, and 7 can be only
partially compared. The Armour chromium was extruded probably below the recrystallization temperature (930°C) while the
Bendix composites were extruded above (1200°C). Secondly, Metcalfe ([3], p. 60) reports a nitrogen content of 3100 ppm in the
Armour chromium at the conclusion of processing which represents two orders of magnitude higher than in the Bendix composites. With this in mind, it is not surprising that the Armour chromium is so strong at temperatures up to 900°C, and no elongation
is measured below 350°C. The strength of the Armour material
is rapidly decreasing as the recrystallization temperature (930°C)
is approached, while the composites exhibit a much more gradual
loss in strength. The ductility of the Armour chromium is decreasing rapidly at 900°C but this trend may reverse at higher
temperatures as in the work reported by Masterson ([1], p. 118).
The increase in strength and decrease in elongation of the Cr—
Nb—MgO composite at 1000°C is probably due to surface reactions
with air during the tests. At higher temperatures, these surface
effects become secondary to the softening of the entire matrix.
The composites show usable tensile properties from room temperature to 1250°C, while the useful range for Armour chromium
does not begin until about 400°C.

Summary and Conclusions

An alternative method for the fabrication of chromium composites has been described in this paper. This method consists
of extrusion of powder compacts and subsequent heat treatments
as opposed to the normally used technique in which the material is
sintered before extrusion. The three compositions included in
this investigation were Cr, Cr—MgO, and Cr—Nb—MgO. A commercial grade of electrolytic chromium powder was used in each.

The following conclusions can be drawn from this investigation:

1. Room temperature ductility has been exhibited by chromium — magnesium oxide composite material which was sintered in either hydrogen or vacuum after hot extrusion of a powder compact.
2. The chromium — magnesium oxide composite had a finer-grained microstructure after sintering at 1600°C than pure chromium processed by the same method.
3. The loss in density observed with increasing sintering temperature can be attributed to internal gas evolution in both the pure chromium and the chromium — magnesium oxide composite. In the chromium — niobium — magnesium oxide composite, the Kirkendall effect supplements this action.
4. The alternate fabrication approach described herein produced weaker, less-dense material than that made by the normal fabrication method.

Acknowledgments

The extrusions were made under the technical supervision of Mr. V. DePierre of the Metals and Ceramics Laboratory of ASD at Wright-Patterson AFB, Ohio. The lattice parameter studies were made by D. M. Scruggs of Research Laboratories Division of The Bendix Corporation. We also gratefully acknowledge the illuminating discussions with G. C. Kuczynski, Professor of Metallurgical Engineering and Material Science, University of Notre Dame and Professor A. W. Allen of the Ceramic Engineering Department, University of Illinois.

References

1. J. F. Masterson, "Development of Chromium Composite Alloy with High Temperature Oxidation and Erosion Resistance," ASD-TDR-63-297 (April, 1963).
2. P. Lowenstein, L. Aronin, and A. Geary, "Hot Extrusion of Metal Powders," in: Powder Metallurgy, Interscience Publishers, Inc., New York (1961), pp. 563-583.
3. A. Metcalfe, S. Spachner, and W. Rostoker, "Mechanical Properties of Chromium," WADC-TR-58-342 (July 1958).
4. D. M. Scruggs and J. F. Masterson, "Development of Chromium Composite Alloy with High Temperature Oxidation and

Erosion Resistance," Progress Report, Contract AF33(657)-8422, Task 738102 (August 1962).

5. K. A. Moon and G. A. Consolazio, "Volume Change and Gas Evolution on Heating Electrolytic Chromium," in: Ductile Chromium, American Society for Metals, Cleveland (1957), pp. 216-228.

Chapter 11. Isostatic Extrusion

Hydrostatic **Extrusion** of Metal Powders*

A. Bobrowsky

Pressure Technology Corporation of America
Woodbridge, New Jersey

The compacting of powders can be done in powder containers with either rigid or flexible walls. A typical powder container with rigid walls might consist of a cylindrical die with closed end, pressurized by a moving plunger entering the other end (unidirectional single-action pressing). The powder can be either in a sheath or loose in the die. The compaction may also occur at elevated temperature, termed hot pressing, in which case a pressing and sintering operation will occur simultaneously.

The compacting of powders is frequently accomplished by the use of fluid pressure exerted on the walls of a flexible container, which is most frequently a flexible sheet of an elastomer or a plastic. In many cases it is possible to insure that the fluid pressure is approximately equal over the outer surface of the contained powder. For reasons that are not clear to the writer, the term

---

* This process, developed by Pressure Technology Corporation of America, Woodbridge, New Jersey, offers most interesting aspects in powder metallurgy. Experimental work has indicated the usefulness of the process; however, no publication showing data of the physical properties of powders compacted by this method has yet been published. [The Editor]

"isostatic" has sprung up, whereas the term "hydrostatic" has been in accepted use in classical mechanics for many decades. One reason given for the term isostatic is that not all liquids are Newtonian liquids, whereby the liquid may support a shear stress. The argument appears to be pedantic, and the term hydrostatic is consequently used in this paper to signify approximately equal pressure exerted on the powder contained in a fluid, especially a liquid.

It has already been shown by H. Hausner [1] that the density of a powder compact increases monotonically with increase in pressure. It would consequently appear that the highest possible compacting pressures should be most helpful in obtaining the highest density compactions.

During the last 20 years there has been an increasing tendency for use of greater compacting pressure, as the availability of suitable equipment has increased.

Pressure Technology Corporation of America has worked with liquid pressures up to 450,000 psi and above. These pressures are employed as environmental pressures in which working and forming of metals is carried out. A major purpose of the high environmental fluid pressure is to permit the material under pressure to become more ductile. There has been considerable experimentation, chiefly by the late Dr. P. W. Bridgman [2], on tensile tests of metals under high pressures. Dr. Bridgman showed that, as pressure increased, the ductility of a tensile specimen became greater [3].

The reasons for the increase in ductility of materials under pressure are not completely known. Bridgman indicated that fracture would not occur so long as the tensile stress in the material to be fractured was lower than the compressive pressure applied by the fluid environment. In the case of ductile materials that neck appreciably during a tensile test, a hydrostatic tension is built up due to the curvature of the neck, and this tension ultimately becomes so great as to exceed the difference of the environmental pressure and the applied tensile stress.

In the case of brittle solids, it is found that their strength rises to a point always equal to or in excess of the applied fluid pressure so long as the material remains brittle. It can also happen that a brittle material becomes ductile under pressure, as for example with tungsten. In this case, necking will occur and the same comments apply as were given above for a ductile specimen.

The essence of the effect of environmental pressure on solids is that ductile materials tend to become more ductile and brittle materials tend to become ductile under high fluid pressure.

The situation with respect to powder particles is somewhat different, inasmuch as powder particles do not have full bearing against other particles when they are pressed together. Portions of the powder particles are unsupported and, consequently, relatively high shear stresses may be generated in the powder particles. The powder particles, however, may support higher nominal stresses than the bulk material would, since the portion of a powder particle that is in contact with another particle is under a locally hydrostatic stress situation, on which are superposed shear stresses. As the distance from the point of contact in the powder particle becomes greater, the hydrostatic component diminishes, but so also does the shear stress. This effect has been termed the "principle of massive support," [4], whereby anvils of solids are found to withstand higher nominal stresses on small contact areas than the bulk solid as a whole can withstand in a tensile or compressive test.

Because of the principle of massive support, it is found that powder particles will flow plastically considerably more due to small areas of contact than would be expected from the behavior of the bulk material in simple tensile tests. On this basis it has been found that high compacting pressures are more efficacious in producing a denser compact than are lower pressures. Fragmentation of powder particles does not readily occur despite the fact that relatively high shear stresses might be thought to be generated by local contact stresses between powder particles.

In Pressure Technology Corporation of America's investigation of the compacting of particles of metal and ceramic powders at pressures up to 450,000 psi, one of the first considerations has been the selection of a sheathing material for the powders to be compacted. Elastomers and plastics may be used so long as they do not tend to become brittle in the high-pressure environments. Further, the material used as a sheath must not react appreciably or at all with the pressurizing liquid. Inasmuch as water solidi-

fies at pressures in the neighborhood of 200,000 psi, it may be
used as pressurizing liquid only for pressures below that level.
Relatively large compacts may be made at such pressures. A
chamber suitable for use at pressures up to 100,000−120,000 psi
can be furnished for commercial use with inside diameters up to
18 in. For pressures above that value, one may approximately
scale down the diameter of the vessel as the pressure increases.
These pressures are conveniently obtained by motion of a piston
in the chamber.

For pressures above about 200,000 psi there are several al-
ternatives regarding the choice of liquid for compaction purposes.
Organic hydrocarbons are frequently suitable for use at pressures
above 200,000 psi and have been found to be useful at least to
450,000 psi. The generating of such pressures is accomplished
industrially by moving a ram into the pressure chamber to compress the liquid. There are currently available methods for rapidly sealing and unsealing pressure chambers, and there are even
pressure chambers that have no permanent seals. The cyclic rate
of operation of such chambers can be relatively high, in the order
of 2500/min for small chambers to 1/3 min for large chambers.
The choice of cyclic rate is dictated by consideration of capital investment. High cyclic rates require equipment to produce high
rates of pumping of liquid which, in turn, necessitates larger and
costlier installations. Hence, if there is no great need for extremely high cyclic rate, one selects a cyclic rate suitable for
use in one's production facilities and with the lowest cyclic rate
compatible with production schedules.

It is frequently desirable to evacuate the powder compact contained in the sheath prior to pressurization, so as to eliminate
pockets of gas that could be trapped in the compact under high
pressures. It can easily be seen that if air is present in a compact that is pressurized to 450,000 psi, the pressure of the gas
contained within the compact may be so great as to rupture bonds
produced between the particles of metal powder of the compact.

The compaction of powder particles under high pressures
sometimes has an aspect of cold welding. This is brought about
because the powder particles may be pressed against each other,
and then slight relative movements may cause cold welding of the
particles. These welds could be broken by entrapped gases unless
the sheath is evacuated.

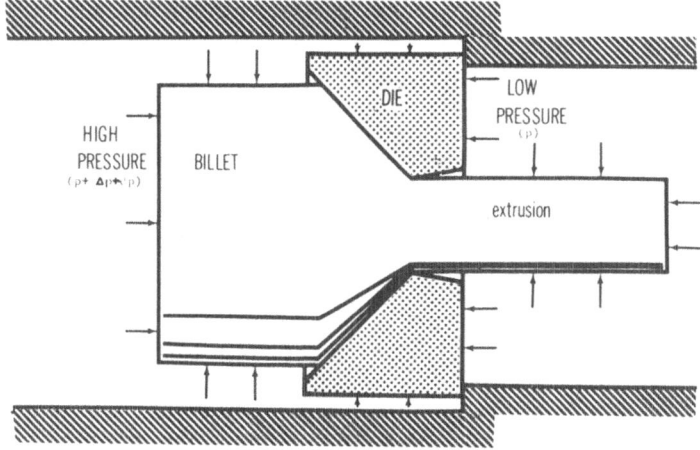

Fig. 1. Schematic drawing of fluid-to-fluid extrusion process. Billet,
die, and extrusion are completely within the pressure environment
throughout the process.

Sheaths have also been made of thin metal tubing. This sometimes offers advantages with respect to the evacuation procedure.
It is important to fill metal sheaths relatively fully or else buck-
ling of the sheath may occur with consequent distortion of the geometric shape of the compact.

It is, however, possible to employ a process termed "fluid-to-
fluid extrusion" for the compacting of powders [5,6]. This process is shown schematically in Fig. 1. The metal sheath containing the powder is termed a billet. High pressure exerted behind
the billet forces it through a die of smaller diameter than the outside diameter of the billet. The resulting extrusion is permitted
to enter a region under lower fluid pressure. It can consequently
be seen that the entire process of reducing the diameter of the
sheath and increasing its length is conducted at an environmental
pressure no lower than the receiver pressure during fluid-to-
fluid extrusion. Examples of such fluid-to-fluid extruded sheaths
are shown in Fig. 2.

The benefits to be gained by the use of extremely high pressures have not yet been explored in great detail (so far as open
publication is concerned). It is reasonable to assume, however,
that proprietary investigations have been made along these lines.
It is also reasonable to assume that such investigations may have

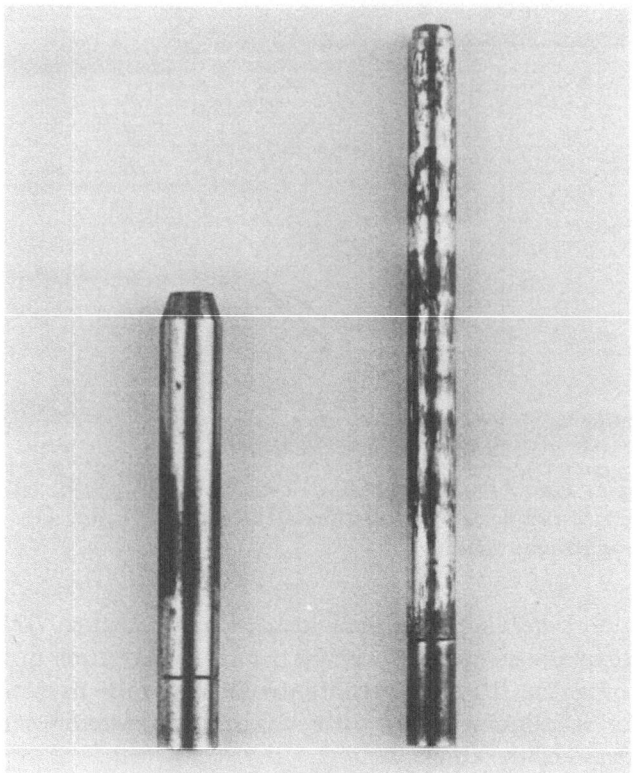

Fig. 2. Fluid-to-fluid extruded steel sheath (a); original sheath
(b) — evacuated.

led to higher density of compaction under high environmental fluid
pressure.

    The question of operational safety sometimes arises. The
safety hazard is minimized and is essentially nonexistent with
proper pressure equipment and process procedures. One key to
eliminating the safety hazard is to minimize the volume of fluid
employed. This is usually done by using "dummy metal" to fill
space in the pressure container that would otherwise be filled by
liquid alone. In this way, the amount of liquid is minimized, as
is the potential energy contained in the liquid.

Further, it is desirable to use liquid of the smallest compressibility; again, in order to minimize potential energy contained in the liquid due to pressure. Water is one of the least compressible of the common liquids, although liquid metal such as mercury is even less compressible. The use of water as a pressurizing medium is consequently relatively desirable.

It has been found that if one surrounds the pressure chamber by a steel pipe or by a portable shell, one obtains a fairly large energy absorption in the shell, in the event of catastrophic failure. Another technique is to have a soft iron shell as an integral part of the pressure chamber, that ordinarily is under no stress. If the pressure chamber should rupture, the soft iron shell stretches slightly to absorb the energy contained in the pressurized liquid.

It should be emphasized, however, that proper design of pressure chambers should lead to absolutely no safety hazard. This is in contrast to views held by people inexperienced in high-pressure work. The author, in 18 years of high-pressure engineering experience, has yet to observe a catastrophic failure. In general, a pressure chamber should last forever. In the most unlikely event of cracking of the pressure chamber, there should occur at most a cracking sound and an immediate drop of pressure in the chamber with no outward manifestation of part failure. The situation can be different if gas is used as the pressurizing fluid. The writer, however, consequently recommends the use only of liquids as pressurizing media. It is sometimes beneficial to induce a failure, say, at 300,000 psi pressure by precracking a pressure chamber in order to demonstrate the lack of hazard if such a failure were to occur.

Properly designed pressure equipment is engineered so that shear stresses are minimized. This is accomplished by the addition of compressive components of loading to pressurized components, so that there is a high hydrostatic compressive component in the material (which will not produce failure), and a relatively low shear stress, preferably below the endurance limit of the pressure chamber. It goes without saying that all other members of a pressure-chamber assembly, such as tie bolts, etc., are loaded so lightly as never to be subject to failure.

The first applications of high-pressure compacting for a commercial group that is considering entering the field will usually be with small compacts. This generates both confidence in equipment and knowledge of the improvement of physical properties communicated by the use of the high environmental pressures. Once these two points have been established, there is no bar to the use of as large chambers as can be usefully employed.

References

1. H. H. Hausner, Powder Metallurgy – Principles and Methods, Chemical Publishing Co., Inc., New York (1947).
2. P. W. Bridgman, The Physics of High Pressures, G. Bell, London (1949).
3. P. W. Bridgman, Studies in Large Plastic Flow and Fracture, McGraw-Hill Book Company, New York (1952).
4. R. H. Wentorf, Modern Very High Pressure Techniques, Butterworths, London (1962).
5. A. Bobrowsky and E. A. Stack, "Fluid Extrusion as a Drawing Process," Nature, 198(4879):474-475 (1963).
6. A. Bobrowsky et al., "Extrusion and Drawing Using High Pressure Hydraulics," ASTME Paper SP 65-33 (November 1964).

Chapter 12 . Powder Forging

Beryllium Powder Forging

N. G. Orrell
Wyman–Gordon Co.
Worcester, Mass.

During the late 1950's considerable interest was developing in beryllium's unique qualities and its potential in new applications. At this same time, the Wyman-Gordon Company was exploring new areas in which to utilize its large presses. The limitations of the press-sintering technique for beryllium were evident, and this, coupled with a consideration of the commercial powder business, led to the development of a new method of beryllium fabrication.

It was envisioned that with high unit p r e s s u r e s of 20,000-100,000 psi and with conventional forge heating techniques the present two-phase processes of compaction and sintering could be married into a single hot-forging cycle. In addition, actual metal movement and deformation could be accomplished as in conventional forging, and it was anticipated that this would impart the typical forging qualities of a uniform and well-worked structure.

The first successful forgings were produced in January 1959; by early 1960 beryllium forgings were being shipped in production lots. Since then, Wyman-Gordon has forged several thousand beryllium components in 18 configurations.

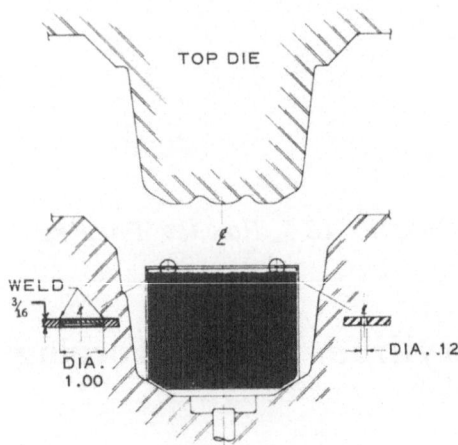

Fig. 1. Details of die and powder can.

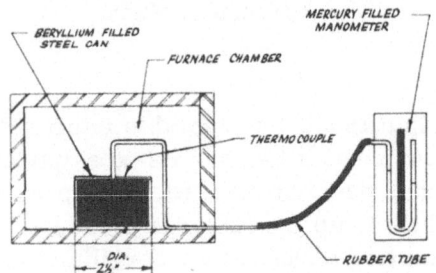

Fig. 2. Schematic diagram of equipment used in
tests.

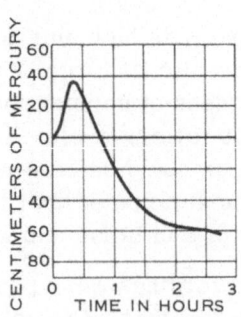

Fig. 3. Pressure versus time
during the outgassing process.

The size of powder forgings has ranged from less than 5 lb to as large as 1200 lb. Diameters have ranged from 3 in. to 60 in., as-forged thickness from less than 0.5 in. to 12 in., and projected area from 20 in.^{2} to 2800 in.^{2}. The configurations produced have been many and varied; shapes for structural use include caps, cones, and cylinders.

During this period an appreciably greater understanding of this metal and its characteristics was gained, and the cost of forged products was reduced by about 50%. Shapes which earlier were considered impossible have been produced, and revised and improved processing techniques have provided increased tensile property levels on the order of 80,000 psi UTS with as high as 12% elongation.

The Basic Beryllium Powder Forging Process

Many variations may be used to produce a final product: forging press-sintered material, forging powder, and forging directly from ingot material. The variations also include subsequent, or reforge, operations in any of the major approaches and both the jacketed and bare-forging techniques.

The basic powder-forging process commences with the can and the powder (Fig. 1). The can design varies according to part shape; the can is basically a light-gauge welded container of steel. The powder used is usually the standard 200-mesh grade, delivered in 25-lb lots packaged in plastic bags for ease in handling. All powder is 100% inspected by x ray to ensure cleanliness. Additional tests are carried out if the powder quality is suspect. The chemistry requirements are based on customer specifications for the final part; BeO content and subsequent Be assay can be varied to meet numerous conditions. The sieve analysis, or particle-size distribution, is determined for each incoming lot of powder.

Acceptable powder is then loaded into special overhead hoppers or prebagged to the desired weight for use in making the parts. At this point the cans have been assembled by welding, provided with loading and vent ports, numbered serially, sandblasted, and vacuum-cleaned. The cans are then placed into special filling fixtures under the hoppers and are simultaneously vibrated and filled with the correct weight of powder.

After filling, the loading ports are plugged and welded shut, and the cans moved to the forge shop. Special handling devices are employed throughout the processing cycle; cans are kept upright during the filling, transferring, and heating stages, primarily to prevent disruption of the initial packed condition and to allow for proper outgassing. The powder cans are placed in a typical gas-fired box furnace on fixtures and preheated to 1000°F (Fig. 2). The time cycle is long enough to allow the inner mass to reach this temperature (Fig. 3). When all outgassing has taken place and the container is essentially at atmospheric pressure, each can is removed from the preheat fire and the vents are welded shut. Each part is then transferred to a rotary furnace at 1800°F and held for 4-5 hr.

When at temperature, the heated parts are removed and hydraulically press-forged in one pass to final shape. Again, handling is very critical, since any damage to the powder package at this point can result in a defective final part.

Press forging is accomplished on one of a number of hydraulic presses depending upon part size and shape. The selection is usually based on a desire to have available unit pressure equal to about 25,000 psi, as it appears that at least this amount is required to ensure full density of the part. Filling the cavity is also important to ensure full density, and good fill is usually ensured by providing the proper exit restriction in the die design. Correct location of the blank in the die is important to prevent offside deformation which can use an underfill, and when designing the dies good mating surfaces are provided to center the work piece. It has also been found important to use caution during the actual press stroke. Parts are formed with the press speed control set at an advance of about 30 in./min. Care is taken to be sure that a steady, constant stroke is employed and the upper die is not slammed into the start piece.

Lubrication is by conventional spraying of the die cavities with an oil and graphite compound. In some cases, the dies are periodically water-cooled between pieces to prevent heat buildup in the tools and subsequent seizing of the blocks.

The usual stamped serial number cannot be employed on the sensitive beryllium components; instead, a stamped tag, to be carried with the forging throughout the remainder of the cycle, is placed on each part.

It should be noted that throughout the entire cycle being described, the plant safety personnel are on hand with equipment to monitor air pollution and to administer any desired precautionary measures.

After the forging operation, all parts are conventionally air-cooled. Care is taken to assure that the cooling is not too rapid.

At this point, the final processing, inspecting, and evaluating of finished components begins. The jacketed forged part is first cleaned by sandblasting. Next, the can is removed either by pickling, machining, or mechanical means. The selection of the method is dependent upon the shape and size of the component, and upon the amount of bonding between can and beryllium.

Following the can removal, the necessary machining to remove integral test material and chemistry chip samples is accomplished. The part is then machined to final dimensions, either for a rough contour or a finished design, depending upon the requirements. In all cases, at least a rough cleaning all over is provided to permit an adequate evaluation of final part quality. Inspection methods such as Zyglo and x-ray require machined surface finish to obtain good interpretation of conditions.

The machining problems have been typical general machine-shop problems. With good practice, beryllium is readily machined to close tolerances with a good surface finish.

Designing Criteria and Limitations

The design of a forged beryllium part, as with most materials, begins with a study of the part required. Rarely with beryllium does a user offer a forging design, because users have little knowledge of the forge-vendor's capability.

The final, or rough-machined, part is laid out full size, and consideration is given to applying a forged envelope around it. Many materials, including beryllium, are forged with the intention of machining all over to yield a final part; the prime reasons for this are the tight dimensional tolerances required on many com-

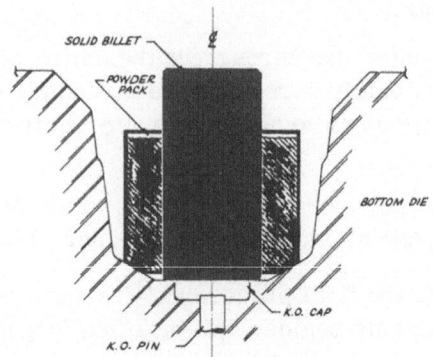

Fig. 4. Comparable initial shapes.

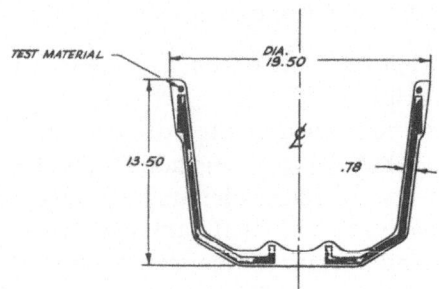

Fig. 5. Final forged shape of beryllium part.

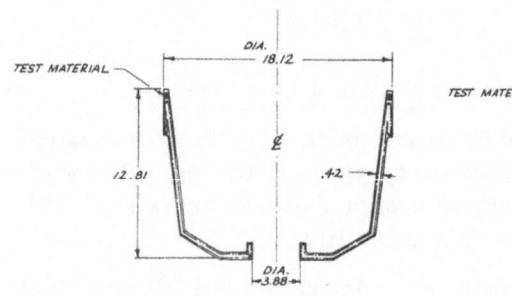

Fig. 6. Rough-machined envelope.

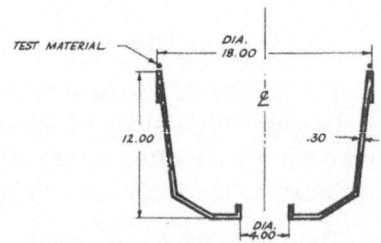

Fig. 7. Sample finished machined beryllium part.

ponents and the need for absolutely clean surfaces for final part application. When designing the forged envelope, the forging engineer is concerned with the extent of added envelope coverage, as this can greatly increase the required input and hence the cost. To be taken into account are such items as the expected extent of surface contamination, expected frequency, location and depth of surface imperfections such as ruptures, and control of the physical dimensions of the part (Figs. 4-7).

We must remember another variable relating to beryllium and the powder method of forging which affects this coverage. In powder forging, a canned mass of powder is deformed and the flow of both metal and can must be anticipated. Throughout most of the deformation operation the can material is the most difficult to flow, and hence is somewhat unpredictable, as both thinning and gathering can occur. Thus, final forge shape is generated by the cavity plus the amount of can material at any given location.

After the envelope has been laid out, the blending of all surfaces and sections into an integral mass must be carried out, and such factors as corner and fillet radii, necessary draft conditions, and minimum panel and wall thicknesses have to be determined and applied.

At this point, the final die contour must be considered and thought must be given to two aspects foreign to typical die forging. Traditionally, the finish die is generated to the exact finish forged design of the part, with allowances for shrink and provisions for saddle, gutter, and locks. In beryllium powder forging, in addition, allowance must now be made for the can material which will lie as a barrier between the part and the tool. Many times this cannot be prejudged and changes must be incorporated after a trial run.

Selection of a starting billet size in typical forging presents no problem, as its volume is usually only slightly larger than that of the die cavity. In powder forging, it is necessary to plan the starting powder package, and because this can only be packed to 50-55% of density, it will be approximately twice the volume of the final closed shape. In die shapes, the accommodation of twice the final volume is quite difficult, and in some shapes it is impossible without an entire change in design concept. The usual ap-

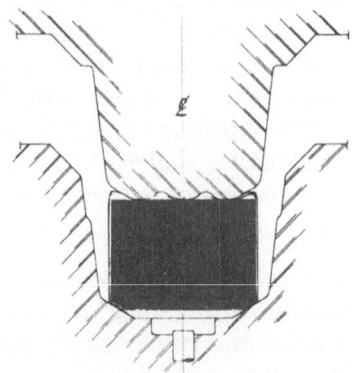

Fig. 8. Beryllium forging sequence:
initial contact.

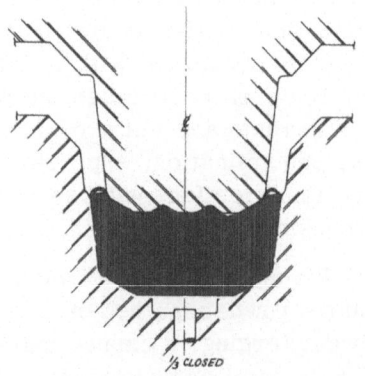

Fig. 9. Beryllium forging sequence:
one third of stroke.

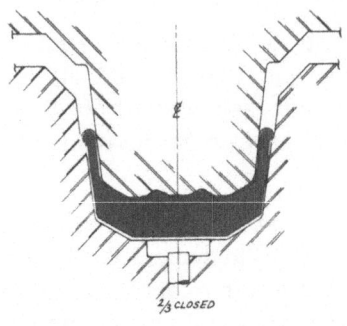

Fig. 10. Beryllium forging sequence:
two thirds of stroke.

proach is to employ a pot type or fully enclosed die such as would be used on a back extrusion.

The purposes of a can are fivefold: (1) the basic purpose is to house the powder so that normal forging die-tolerances can be employed and to permit the handling of specific quantities of powder during heating and prior to consolidation; (2) a can provides an effective barrier to the toxic oxide and dust of beryllium during the earlier phases of the cycle; (3) canning provides an effective barrier to external elements that would contaminate the forging material during hot-working; (4) in powder forging the can is designed to provide resistance to metal flow to ensure gradual and complete densification of the powder; (5) the container provides an effective heat barrier insulating the work piece from the tools.

A typical can today is usually of a simple geometric shape resembling a billet, assembled from 1020 steel sheet by welding, and kept as thin as possible, usually $\frac{1}{8}$-$\frac{3}{16}$ in. thick. Loading

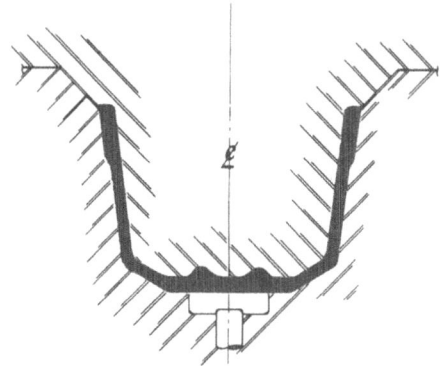

Fig. 11. Beryllium forging sequence: die to die.

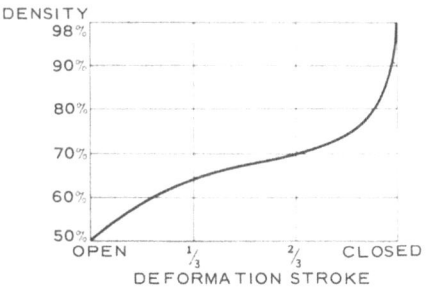

Fig. 12. Relative action during forging stroke.

ports are closed by thin flat discs welded into a recess after filling, and the vents are simple drilled holes.

Figures 8-11 illustrate the action and reaction that takes
place during the pressing operation. It should be pointed out that
during this cycle the powder material is both densified and deformed to final shape.

Although experience in the powder forging of beryllium has
still not reached the stage of sophistication of wrought bar materials, the powder forging technique offers considerable savings
in material utilization over machining from press-sintered block.
Table I shows this graphically. Further, in powder forging, the
entire thermal-pressure cycle is carried out in a few seconds to
accomplish a complete transition from powder mass to consolidated and well-worked final shape (Fig. 12).

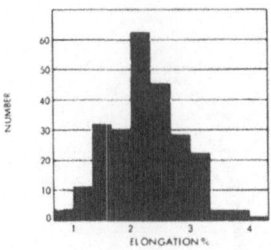

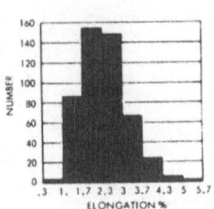

Fig. 13a. Forged beryllium tensile properties; frequency distribution for 241 parts; W-G 19033; long transverse direction (see also Figs. 13b, 13c): elongation.

Fig. 14a. Forged beryllium tensile properties; frequency distribution for 516 parts; W-G 19026; long transverse direction (see also Figs. 14b, 14c): elongation.

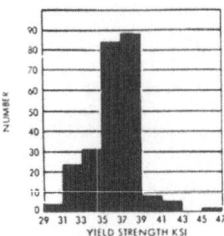

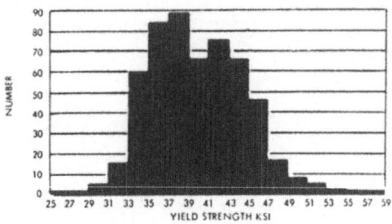

Fig. 13b. Yield strength (241 parts; see Fig. 13a).

Fig. 14b. Yield strength (516 parts; see Fig. 14a).

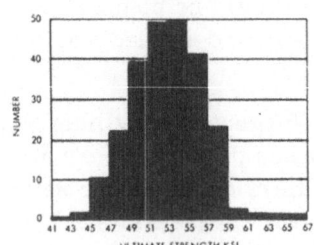

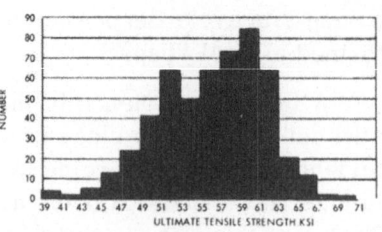

Fig. 13c. Ultimate strength (241 parts; see Fig. 13a).

Fig. 14c. Ultimate tensile strength (516 parts; see Fig. 14a).

Table I.   Relative Required Beryllium Input Weight
Press-Sintered Block Versus Powder Forgings

Part	Block required (pounds)	Powder required (pounds)
Brake disc., 8 in. dia.	2.6	2.4
Guidance comp., 6 in. dia.	2.9	2.0
Structural beam, 28 in. long	197	88
Beryllium cap, 20 in. dia.	384	58
Missile spacer, 30 in. dia.	580	285
Rocket fairing, 72 in. dia.	8,400	700

The factors that have limited the shape sophistication achieved in beryllium powder forging are the lack of experience, high dollar risk, small quantities, and short delivery requirements. Parts with deep pockets, thin sections, tight fillets, and severe irregularities are not a typical product today.   Section thickness is limited by the combination of rapid chilling of the thin sections and the brittleness of the material.

Recent work has shown that the real problem in forging or reforging solid material is the inherent lack of ductility and the resulting limited amount of work that can be accomplished per pass.

The only answer to the whole competitive problem of forging will arise with the requirement for a superior product and the realization that a forging meets this need.

Mechanical Property Capability

As might be expected, the mechanical properties of forged beryllium powder are generally at a level intermediate between those of parts produced by press sintering and those shaped by solid forging.   They will also vary over quite a wide range, depending on the severity of the working.

Statistical summaries of properties obtained from some 241 parts of one configuration (Figs. 13a-13c) and 516 parts of another configuration (Figs. 14a-14c) are reproduced here.   The results are for an integral test from each forging and represent several variations in forging practice throughout the production cycle. The tests are from the location in the part which has been found to give the lowest properties; thus these results are representative of the

FORGED BERYLLIUM BOWL

(From Powder)

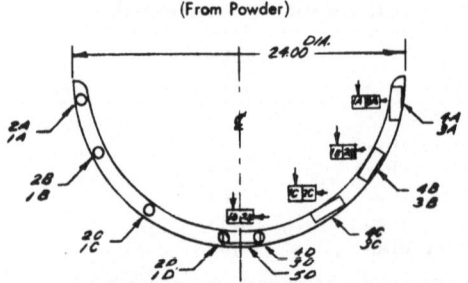

Test Location	Test Direction	Yield Strength psi 0.2% Offset	Ultimate Strength psi	Elongation % in 4 D (Scribe)	Elongation % in 4 D (Extensometer)
1A	Tangential	39,900	56,100	4.0	3.9
1B	T	37,700	61,000	6.3	7.0
1C	T	36,900	59,700	6.9	6.2
1D	T	37,300	59,700	4.5	6.1
2A	T	38,900	58,100	4.5	4.9
2B	T	37,900	59,900	5.3	5.7
2C	T	35,000	59,300	5.4	6.0
2D	T	37,300	62,200	6.3	6.9
Average Tangential		37,600	59,500	5.4	
3A	Radial	42,200	68,200	9.7	7.4
3B	R	39,100	63,000	7.3	7.6
3C	R	38,700	62,200	8.5	7.6
3D	R	37,700	61,800	6.7	7.2
4A	R	40,800	65,600	7.3	7.5
4B	R	37,700	61,200	5.0	6.5
4C	R	37,300	61,400	7.5	7.4
4D	R	37,300	61,400	6.9	6.9
Pole		36,900	61,600	7.0	7.6
Average Radial		38,800	63,100	7.4	
General Average		38,200	61,300	6.4	6.6
Specification		30,000	40,000	1.0	

Fig. 15. Data sheet #1.

worst condition in the parts. None of these parts were given special heat treatments after forging to improve properties; most
were air-cooled after forging, while the balance were furnace-
cooled.

The six data sheets (Figs. 15-20) were prepared to illustrate
the properties obtained on various shapes and sizes of parts. The
first sheet (Fig. 15) shows the test locations in a typical bowl shape

ELEVATED TEMPERATURE TENSILE RESULTS ON 24" Ø BOWL

All tests were taken in the tangential direction from the forging sidewall.
The tensiles were held at temperature for ten minutes before being pulled.
The tensile diameters were .250".

Test Temperature	0.2% Yield Strength	Ultimate Tensile Strength	Elongation %
70° F - Average	37.7	59.0	5.4
200° F	33.8	54.6	10.2
410° F	28.5	43.8	19.0
600° F	25.5	35.7	25.1
800° F	21.0	32.6	18.1
1000° F	19.6	27.3	17.4
1200° F	13.9	18.4	6.9
1400° F	5.7	6.1	2.8

Fig. 16. Data sheet #2.

FORGED BERYLLIUM FLARE

(From Powder)

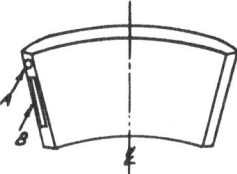

Typical tensile properties would be expected to fall in the following ranges:

Location	.2% Yield Strength psi	Ultimate Tensile Strength psi	Elongation %
A	45,000	68,000	3.5
	35,000	50,000	2.5
B	45,000	72,000	8.0
	33,000	57,000	3.0

Fig. 17. Data sheet #3.

FORGED BERYLLIUM CYLINDERS
HOLLOW EXTRUSION - BACKWARD

(From Powder)

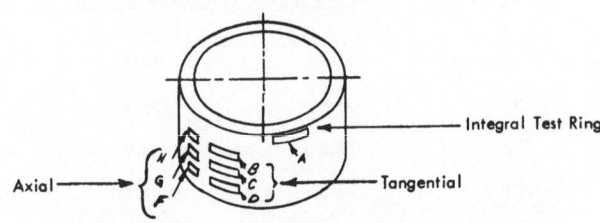

Specimen	.2% Yield Strength	Ultimate Tensile Strength	Elongation %
Part #1 A	42,000	62,200	4.1
B1	44,300	68,600	5.6
B2	43,700	66,600	5.0
C1	44,800	65,800	4.3
C2	44,700	65,000	3.7
D1	44,800	71,600	6.0
D2	43,600	62,600	4.0
F1	42,900	71,700	8.0
F2	43,900	72,400	7.5
G1	43,300	71,500	7.1
G2	43,700	72,400	7.8
H1	45,100	72,000	6.6
H2	43,900	71,000	7.7
Part #2 B1	45,000	62,400	4.9
B2	43,800	59,700	3.6
D1	44,400	57,500	3.0
D2	43,800	56,500	3.3
F1	45,200	62,800	4.1
F2	44,100	62,800	4.8
H1	44,000	76,000	6.4
H2	43,900	76,700	6.1
Specification	30,000	40,000	1.0

Fig. 18. Data sheet #4.

FORGED BERYLLIUM CYLINDER
(Backward Extrusion from Powder)

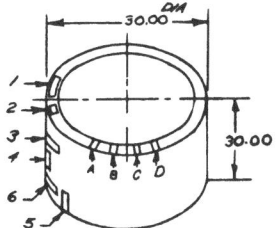

Reported Properties:

Location	0.2% Yield Strength	Ultimate Tensile Strength	Elongation %	R. A.
1	41,800	55,000	1.7	---
2	41,200	58,900	2.8	---
3	39,700	55,000	2.9	---
4	39,700	63,000	5.5	---
5	39,700	55,400	3.8	---
6	40,100	62,200	5.5	---
Specification	30,000	40,000	1.0	
A	*	43,500	0.4	4.0
B	38,900	40,800	0.2	4.0
C	40,200	42,600	0.8	4.2
D	44,500	50,600	0.7	1.5

* Strain gage failed before yield.

Fig. 19. Data sheet #5.

configuration about 24 in. in diameter with a wall thickness of approximately $\frac{3}{4}$ in. Tests were taken in radial and tangential directions throughout the bowl. This figure also gives the properties obtained in the two directions. The second data sheet (Fig.
16) shows elevated-temperature tensile results taken from one of
these 24-in. round bowls. The third data sheet (Fig. 17) shows
properties in two locations from another typical powder forging.
The fourth sheet (Fig. 18) shows schematically the test locations
in a large cylindrical forging. Room-temperature tensile properties for two different locations in these forgings are also shown.

FORGED BERYLLIUM PANCAKE
(From Powder)

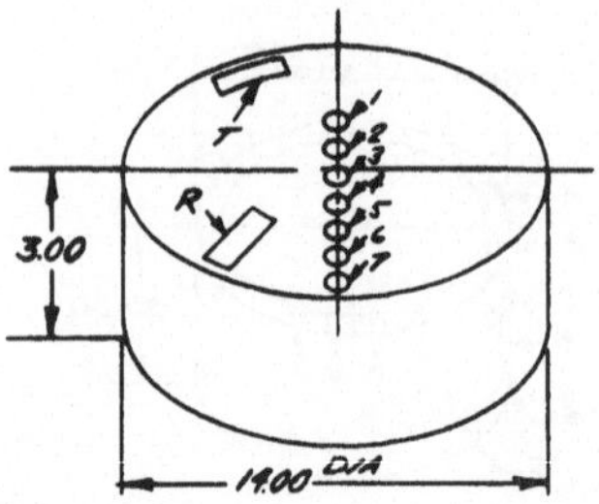

Reported Properties:

Location	0.2% Yield Strength	Ultimate Tensile Strength	Elongation %
R	36,800	52,600	2.3
T	36,200	53,800	2.5
1*	38,600	43,500	0.6
2*	38,600	43,000	0.6
3*	39,300	39,900	0.4
4*	38,300	42,600	0.6
5*	38,800	43,000	0.6
6*	39,100	43,400	0.6
7*	39,200	43,700	0.6

* Short Transverse

Fig. 20. Data sheet # 6.

Since these are different locations in two different forgings, the results cannot be compared directly. However, they do give some indication of property variation from part to part and location to location.

As might be expected, beryllium properties vary with the degree of preferred orientation or with test direction and degree of working in the forging. One example of this is shown in the fifth data sheet (Fig. 19). The lowest ductilities were found in a direction perpendicular to the major direction of metal movement (short transverse direction). The ultimate strength was also lower in this direction but the yield strength was practically unaffected. Another example of the effect of test direction is shown in the sixth data sheet (Fig. 20). Although the tensile properties in this part are not particularly high, it is noteworthy that they are very reproducible from one location to another in the same test direction.

It is apparent that in a relatively few years the state-of-the-art in beryllium fabrication has come a long way. It should be equally apparent that much remains to be understood and accomplished in this field. Some limited experimental work indicates that certain advances offer great potential for improvements in forged beryllium products. It is expected that future efforts in reforging cycles and bare forging techniques will permit the manufacture of sophisticated shapes commensurate with the present capability in typical die-forging of other materials.

Further investigation is needed in sub-sieve particle size and distribution as it affects forgeability and properties. Surface roughness of the particles in relation to compaction and forgeability has not been thoroughly investigated and merits some future attention.

Only a limited amount of work has been carried out at Wyman-Gordon with improved thermal cycles during forging, and no real effort has been put into heat-treating studies. The reason for this has been that improved properties have not been required by the major market served to date. It is hoped to expend more effort in this direction, and it is certain that investigation in these areas can improve part quality.

There remains much to be done to extract a greater proportion of the available advantages from a renovation of the method of producing the metal in conjunction with forging. Increased purity resulting from less interstitials and oxide film will probably improve the forgeability of beryllium and possibly show improvements in tensile ductility, with some sacrifice in strength. The lowering of metallic contamination will have the effect of improving part soundness and increasing reliability.

Compositional changes or alloying may offer potential improvements. In the powder area, a so-called "instrument grade" beryllium is being developed which has a substantially increased BeO level and is reported to give higher tensile properties. It is

of a much finer mesh size than normal and therefore will considerably aggravate the handling problems for powder users. Additional effort is required to make this material and its conversion to
final parts a reality.

It is believed that strong efforts are needed in the area of
producing a beryllium metal product in solid form that can be
wrought. There can be little doubt that the direct conversion of
vacuum-melted ingot to finished part can be valuable from both
cost and quality standpoints. There is a definite need to concurrently realize a substantial scale-up in size from the current ingot product.

An additional suggestion for future emphasis would be that the
current material efforts be complemented by more programs aimed
at component development. There appears to be no better way
to evaluate and prove a material product than to employ it in finished part manufacture.

Slip Casting of Metal Powders

Henry H. Hausner

Adjunct Professor
Polytechnic Institute of Brooklyn
and Consulting Engineer
New York, New York

1. It is a well-known fact that conventional powder metallurgy, i.e., pressure-compacting of metal powders and subsequent sintering of the compact, has several severe disadvantages, such as:

   a. It is somewhat restricted to the production of rather small parts due to the capacity limitations of conventional presses, and due to the relatively high die cost.
   b. It does not permit pressing of certain complicated shapes due to the nonuniform pressure distribution in a mass of metal powders which is caused by the friction conditions prevailing during the application of pressure.
   c. It is economic only if applied to a relatively large-scale production, on account of the relatively high cost of the hardened steel dies.

These are the shortcomings which restrict the application of powder metallurgy methods, sometimes even in cases where a powder metallurgy product would be desirable on account of its special properties.

2. The above is true also in ceramics, when pressure molding is applied. The ceramist, however, has developed other methods for forming ceramic bodies, which eliminate the application of pressure – and I refer especially to the ceramic slip-casting process, known and successfully applied for many years for the production of ceramic goods of large dimensions or complicated shapes. In ceramics a slip or slurry is defined as a fluid suspension of ceramic powders in liquids, and slip casting refers to the filling of a plaster of Paris mold, a negative of the desired shape, with this slip. By capillary forces of the pores, the liquid penetrates into the plaster mold and the ceramic powder particles adhere strongly during the drying operation in the mold. The slip for casting is prepared from water, ceramic powders such as clay and talc, and a small amount of a deflocculant is added in order to prevent the settling of the powder particles, and to create the desirable viscosity of the slip. After partial drying in the mold and final drying in air at room or elevated temperature, the molded ceramic mass is fired (Fig. 1 illustrates this process).

3. This well-known ceramic process of slip casting can also be applied to certain metal powders, and parts of large dimensions or complicated shapes can be molded in this way. The process is characterized by the following steps:

a. Select the slip mixture. Variables are as follows: (1) powder particle size ranges; (2) type and amount of deflocculant; (3) water:metal ratio; (4) viscosity; (5) pH value.
b. Weigh the specific amount of powder. If more than one powder fraction is used in a slip, weigh the individual fractions and mix them.
c. Add the specific amount of deflocculant to the dry powder and blend.
d. Slowly add the appropriate amount of water, stirring mechanically while the water is being mixed.
e. Continue mechanical stirring until a smooth slip is observed and determine the pH value.
f. Alter the pH value to the selected one by adding either concentrated acid or hydroxide solution to the slip by constant checking by pH-meter.

g. Allow the slip to set for a few hours to assure the removal of any air bubbles which may have been formed during stirring, or degas in vacuum.

h. Apply a protective coating to the plaster of Paris mold.

i. Pour the slip slowly into the plaster of Paris mold of the desired shape and size.

j. Allow the slip to set for 10-20 hr in the mold, so that the water can penetrate into the mold material.

k. Carefully remove the metal powder slip casting from the mold and trim away any excess flash material.

l. Dry the metal powder slip casting in an oven.

m. Sinter at the desired temperature and time in order to obtain desired properties of the sintered material.

4. This brief description of the metal powder slip-casting processing steps already indicates that many variables are involved in the process which were never before considered by the powder metallurgist. Some of these variables are:

a. density of the liquid

b. type and amount of deflocculant

c. ratio of liquid to solid

d. viscosity of the slip

e. pH value of the slip

f. reaction of the metal powder with the deflocculant and the liquid

g. temperature of the slip

h. amount of entrapped air

i. material of the mold

j. moisture content of the mold

k. porosity of the mold material

l. rate of drying in the mold

m. rate of drying after elimination from the mold

It should be mentioned also that the selection of metal powders with respect to particle density, shape, and size has to be made much more carefully for slip casting than for conventional pressure-molding. However, some of the variables which usually complicate powder metallurgy are missing in the slip-casting process. All the variables with respect to pressure, pressure dis-

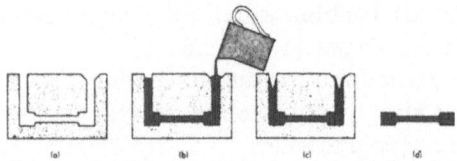

Fig. 1. Principles of slip casting: a) assembled plaster mold; b) filling the mold; c) absorbing water from the slip; d) finished piece, removed from the mold and trimmed. (according to F. H. Norton.)

Fig. 2. Plaster of Paris molds for rectangular and cylindrical parts (courtesy Sylvania-Corning Nuclear Corp.).

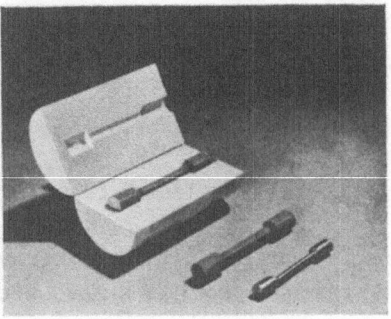

Fig. 3. Plaster of Paris molds for tensile bars (courtesy Sylvania-Corning Nuclear Corp.).

tribution, pressure gradient, and pressing temperature are not at all pertinent in slip casting.

5. The molds used in casting of metal powder slips were made by conventional techniques as used in standard ceramic slip casting. The three main variables controlling the behavior of slip-casting molds are: (1) the plaster—water ratio; (2) the grade or degree of refinement of the plaster; and, (3) the area from which the plaster manufacturer obtains his raw material. Low plaster-to-water ratios produce softer, more absorbent molds while high ratios produce harder, less absorbent material. The source or area, from which the gypsum is mined, influences the properties of the plaster of Paris (manufactured by heating). The degree of particle-size refinement, purification, and control variables in the dehydration process also effect different behavior in the final product.

Figures 2 and 3 show some of the plaster of Paris molds used for the metal powder slip process.

6. Before discussing slip casting further, it might be worthwhile to compare the properties of an individual powder particle in a mass of powder with application of pressure (conventional powder metallurgy) and without the application of pressure (loose powder or slip casting), and to discuss briefly the effects of these properties on the sinterability of the powder. Table I shows such a comparison.

One will easily recognize that the rate of sintering and the material movement during sintering of the undeformed powder particles, such as used in the slip-casting process, will be different from the sintering behavior of the powder particles deformed during pressure compacting. It is quite understandable, as will be shown below, that the grain structure of a loose or slip-cast metal powder after sintering is different from the structure of similar, but pressure-compacted and sintered, powder.

7. The metal powder particle in the slip is freely suspended in the liquid and its movement is caused by the following possible mechanisms:

    a. a vertical movement due to gravity forces

Table I.   Comparison of Properties of Metal Powder Particles Loose (Uncompacted) and Compacted

Loose powder	Compacted powder
1. Packing of particles by gravity forces only results in a loose packing	1. Packing under pressure which overcomes adhesion and friction between particles results in a closer packing system
2. Powder particles keep their original shape	2. Powder particles deform plastically under pressure and lose their original shape
3. Contacts between particles are rather of point-form	3. During deformation the original contact points between particles develop to larger contact areas
4. Powder particles touch each other at ambient temperature	4. Development of heat during deformation, especially at contact areas
5. Particles possess their original surface crystal lattice defects	5. An exchange of atoms between particles takes place at contact areas under pressure, and the types and amount of surface lattice defects changes. On account of these changes and exchanges of atoms, a bond between particles is formed
6. Particles possess their original body crystal lattice defects (point defects)	6. New structural and body lattice defects are formed during plastic deformation (linear defects)
7. Stresses in the powder particles depend on the method of fabrication	7. In addition to original particle stresses, stresses are formed on account of plastic deformation
8. Density of powder mass is low ("apparent" density)	8. Density of powder mass ("green" density) is considerably higher than apparent density, depending on the plasticity of the particles and the applied compacting pressure
9. The shape of voids between particles depends on the powder particle shape (mostly equiaxed pores)	9. The voids are usually elongated in the direction perpendicular to the pressing direction (oriented voids)

b. an irregular movement in all directions, due to the molecular forces of the fluid (Brownian movement)

c. spinning of particles, due to electrical forces

d. movement due to the friction with the liquid during the penetration of the liquid into the mold.

An analysis of these four possibilities for movement of a powder particle indicates that all these movements contribute to a closer packing of the powder particles by movement of the finer particles into the void space between the coarser ones.

8. Among the processing variables mentioned above, the two which are entirely new to the powder metallurgist are viscosity and pH factor of the slip. Both are of utmost importance in the slip-casting process, and greatly affect the properties of the product before and after sintering. These two factors depend, among others, on the liquid-to-metal ratio, and on the type and amount of the deflocculant. In order to produce a useful slip, the suspension of solid particles in the fluid must remain homogeneous and without segregation for a long period of time, despite the settling tendencies according to Stokes' Law. The ratio of specific gravities of the solid to fluid greatly affects the nature of the suspension. As this ratio increases, the settling forces come into effect; they must be balanced by other homogenizing tendencies, such as viscosity of the suspending media, particle size and shape of the suspended particles, and the chemical and electrical relationship between fluid and solid. Defllocculants are used for "thinning the consistency of a slip by addition of an electrolyte."* In ceramics polyvinyl alcohol, gum arabic, various types of alginates, and others are used as defllocculants.

9. The investigations into the slip-casting process described below were made with 316 stainless steel powder received from Charles Hardy Inc. and produced by the Federal-Mogul Research Division of the Bower Bearing Co., Ann Arbor, Michigan. This powder is characterized by the uniformly spherical shape and the high density of the particles. The chemical analysis of the powder is shown in Table II. The screen analysis indicated the par-

---

*"Chemical and Engineering Dictionary," Chemical Publishing Co., New York.

Table II.   Chemical Analysis of Type 316 Stainless Steel Powder

Component	Percentage
C	0.09
Mo	2.00
Si	0.98
P	0.004
S	0.025
Mn	Nil
Ni	10.80
Cr	17.70
Fe	balance

ticle-size distribution given in Table III, which also shows the apparent density of the powder as received and of the individual fractions. For the tests described below, a mixture of two particle-size fractions was used, i.e., 80% 325 mesh fraction was carefully blended with 20% of the 100-150 mesh fraction.

10. At the present extent of investigations with the 80/20 blend of (−325) to (−100 + 150) spherical-type 316 stainless steel and water as the vehicle, the algins have demonstrated the most promise of the aforementioned binders. The two algins used were ammonium alginates and are designated by the trade names Marex and Superloid. These alginates are ammonium salts of alginic acid which, in turn, is a hydrophilic colloidal polymer of anhydro-B-D-mannuronic acid units. Listed below are the properties of these two algins which are important to consider in their application as a slip-casting deflocculant and binder.

	Property		Marex	Superloid
1.	Moisture content		10%	10%
2.	Ash (Ca, Na, chlorides)		4% max	4% max
3.	Particle size		−80 mesh	−20 mesh
4.	pH in aqueous solution		5-6	5-6
5.	Viscosity in centipoises at 25°C in water dispersion	1% binder	85	1200
		2% binder	660	13000
		3% binder	2800	50000
6.	Percent of material in water at which colloidal dispersion exhibits jell structure		7-8%	4-5%
7.	Structure		amorphous	amorphous
8.	Thermal decomposition in the presence of O_2			

All algin products decompose at 210-225°C by carbonizing. Residual carbon can be burned off completely in air at 1250°F. This is the temperature at which the ash content is determined.

Table III. Particle-Size Distribution
and Apparent Density of Type 316 Stainless Steel Powder

Mesh size	Percentage	Apparent density g/cc
as received	-	4.82
− 80+100	trace	-
−100+150	9.78	4.46
−150+200	19.50	4.48
−200+250	4.68	4.48
−250+325	26.95	4.47
−325	39.09	4.39

It should be noted that although the water solution of these alginates is acid, addition of stainless steel powder results in a decidedly basic slip. This might be explained by hypothesizing an oxidation−reduction reaction which results in lowering the ammonium ions in solution with the corresponding result of increasing the pH.

In metal powder slip casting, the deflocculant acts also as a binder for the powder particles. There are, however, good reasons to believe that a chemical reaction takes place between the fluid and the surface of the metal particles, which may affect the sintering process.

11. In ceramics it is well known that a relationship exists between viscosity and the pH value of the slip. To obtain a relationship between viscosity and pH for stainless steel slip casting, six separate batches of metal powder slips were made up of the following constituents:

16.89% −100 + 1500 mesh type 316 spherical stainless steel
67.57% −325 mesh type 316 spherical stainless steel
15.20% distilled water
 0.34% ammonium alginate (Superloid)

After mixing, each of the six slip batches weighed approximately 1200 g.

All slips at this point were checked for variation in viscosities and pH values. All were found to have viscosities (determined on a Brookfield viscosimeter) between 36,000 and 37,000 centipoises, and pH values (obtained with a Beckman pH meter) between 8.6 and 8.7.

It was observed that to adjust the pH to a specific value, small additions of concentrated HNO_3 (to lower the value) and concentrated NaOH solution (to raise the pH) were required. These chemicals were chosen because of the excellent resistance of type 316 stainless steel to attack by them. They were added to the original slip mixtures in drop form. For example, 10 drops ($\sim\frac{1}{2}$ cc) of concentrated HNO_3 were added to one mix to attain a pH value of 7.0. Some lumping of the mix was observed as the acid was added, but this was eliminated by mechanical stirring. Once this pH value was obtained, the batch was set aside for future determination of viscosity and specific gravity and the procedure repeated to adjust the pH of a second slip batch to another specific value. By this method the pH values of the six batches were adjusted to 7.0, 8.0, 9.0, 10.0, 11.5, and 12.7.

The viscosities of each of these adjusted slip mixtures were determined and found to be:

pH	Viscosity, centipoises
7.0	49,000
8.0	42,000
8.6-8.7	36,000-37,000
9.0	27,000
10.0	18,500
11.5	21,500
12.7	56,000

The specific gravity of each was determined and was found to fall in the range of 3.87 ± 0.01 for all slips.

Following the determination of the above data, 10 cc of distilled water was added to each slip mixture to lower the specific gravity and viscosity. This addition of water altered the composition of the slip batches to:

16.75%   −100 + 150 mesh type 316 spherical stainless steel
67.00%   −325 mesh type 316 spherical stainless steel
15.91%   distilled water
 0.34%   ammonium alginate (Superloid)

When the addition of this water was completed by mechanical stirring, the pH value and viscosity of each batch were determined and found to be:

pH	Viscosity, centipoises
7.6	30,500
8.2	26,000
9.0	14,500
9.9	10,700
11.0	11,300
12.7	30,500

The specific gravity of these slip mixtures was found to be in a range of 3.79 ± 0.01.

Again the procedure of adding distilled water to lower the specific gravity was repeated with the addition of 5 cc of distilled water. This addition of water altered the overall composition of each batch to:

16.67%	−100 + 150 mesh type 316 spherical stainless steel
66.69%	−325 mesh type 316 spherical stainless steel
16.30%	distilled water
0.34%	ammonium alginate (Superloid)

The pH values and viscosities of these mixes were redetermined as before and found to be:

pH	Viscosity, centipoises
7.8	25,500
8.3	20,500
8.9	12,400
9.8	8,800
10.7	9,300
12.5	26,500

The specific gravity of these slips was found to fall in the range of 3.72 ± 0.01.

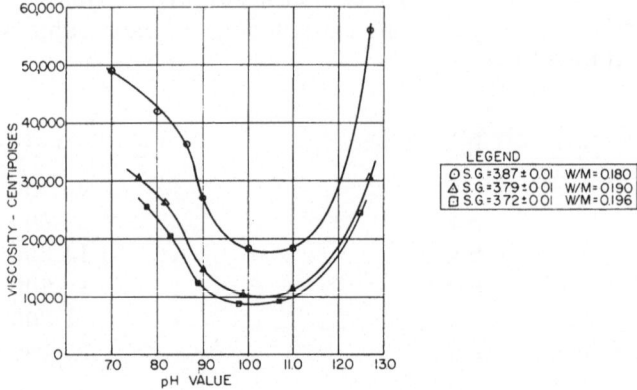

Fig. 4. Correlation between viscosity and pH value
for slip consisting of stainless steel powder particles
in water, containing 0.34% Superloid, for three different water : metal ratios.

The data obtained on the varying specific gravities above were
plotted on a graph (Fig. 4), showing viscosity versus pH as a fam-
ily of curves of varying specific gravity.

Preliminary investigations showed that the decisive factor for
the physical properties of the slip-cast and sintered material is
not the viscosity of the slip, but the pH value. Slips can be produced with the same viscosity, but with different pH values.

Very recently, Dr. Roger C. Bates* of the National Bureau
of Standards stated: "Perhaps the most difficult pH problem fac-
ing the modern analyst concerns the proper use of measurements
in media that are not entirely aqueous. In its strictest sense, pH
has no meaning in such solutions... Yet, in its most general for-
mulation — namely, the operational definition — a pH value exists
for any medium in which the pH cell develops a reproducible emf."

It is not intended here to give a fundamental interpretation of
the above-listed pH measurements, which may be regarded just

―――――――
* "pH and the Modern Analyst," Anal. Chem. 29:15A (May 1957).

as indicators of the chemical composition of the slip and are extremely useful for the precise reproduction of the slip composition especially inasmuch as the viscosity figures of the slip are meaningless as long as they are not connected with the respective readings on the pH meter.

12. There is a strong correlation between the water-to-metal ratio and the amount of the deflocculant in the slip. Increasing the amount of water-to-metal ratio requires increasing the amount of the deflocculant, in order to keep the pH value and viscosity constant. An example is shown in Fig. 5, for Marex and Superloid as deflocculants to keep the pH value constantly at 10 and the viscosity at 1500 cp. This correlation must be carefully watched, inasmuch as it greatly affects the physical properties of the slip-cast product.

13. Preliminary investigation has shown that the density of the cast slips after being taken out from the mold varies with the water-to-metal ratio: the greater the water-to-metal ratio, the less the density. The same is also true for the same material in the dried stage, and to a certain extent also in the as-sintered stage. The variations in density with various water-to-metal ratios are fairly small in the as-cast or as-dried state, and more pronounced in the as-sintered state. An investigation has therefore been made in order to determine the densities at various stages of the process, for 15, 17, and 19% water content.

Slip mixtures were prepared with varying water content while keeping the binder content constant at 0.30% by weight. This variation in water content would necessarily result in a variation in the metal concentration of the slip when the binder content is kept constant. Three slip mixtures were prepared of the compositions shown in Table IV.

These slip mixtures, after mechanical mixing to assure good blending, were adjusted to a uniform pH value of 10.0 by the minute addition of concentrated NaOH solution. The specific gravities were determined on these constant binder-constant pH slip mixtures. The slip mixtures were then cast into $\frac{1}{2}$-in.-diameter by 2.00-in.-long specimens and allowed to set overnight. Following this setting period the specimens were carefully removed from

Table IV.   Effect of Water Content on the Slip Conditions
and the Properties of Test Specimens

Slip composition [1]	Water content %		
	15.0	17.0	19.0
Metal powder[2]	84.70	82.70	80.70
Deflocculant (Superloid)	0.30	0.30	0.30
Water/metal ratio	0.1771	0.2056	0.2354
Specific gravity, g/cc	3.87	3.57	3.37
Specimen conditions			
Density, g/cc, as cast	3.03	4.96	4.88
" as dried[3]	4.77	4.70	4.62
" as sintered[4]	7.18	7.08	6.94
Green strength, psi (as dried)	910	804	743

[1]Adjusted to pH = 10.
[2]316 type SS 80%, 325 mesh, 20% − 100 + 150 mesh.
[3]for 2 hr at 50°C in air.
[4]for 2 hr at 1300°C in dried hydrogen.

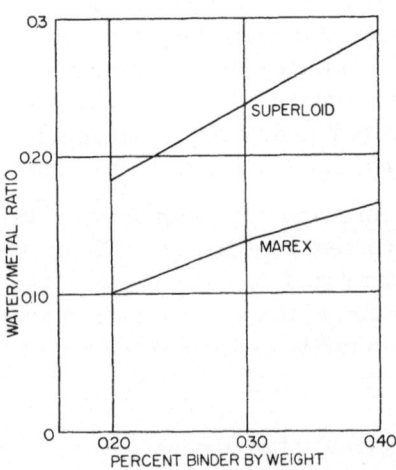

Fig. 5.   The amount of deflocculant (binder)
and its effect on the water : metal ratio (specific gravity of the slip = 3.4, pH value is 10,
and viscosity = 15,000 cp constant for all
water : metal ratios).

the mold and "as-cast" density measurements taken. "As-dried" densities were also determined after a drying period of 3 hr at 50°C in an air oven. As-sintered densities were determined after a 2-hr sintering at 1300°C in a dry hydrogen atmosphere. All density measurements are tabulated also in Table IV. It may be observed that as the water-to-metal ratio increases, the densities tend to decrease.

Table IV also gives the green breaking strength of the as-dried specimen. Experience has shown that, with increasing amount of the deflocculant, the green strength increases and, for example, a slip-cast 316 stainless steel with 0.4% Marex as deflocculant showed a green strength of 1155 psi, which strength permits safe handling of the green compact in any production. There is no doubt that the deflocculant also acts as a binder; however, it is not yet known whether a chemical reaction between the deflocculant and the powder particle takes place, and this will be the subject of a further investigation.

14. The slip-casting process has been used to produce various shapes from stainless steel – complicated shapes such as turbine blades and especially hollow turbine blades with a certain porosity for cooling purposes – as well as flat plates of relatively large dimensions. Plates of 2 in. × 12 in. × $\frac{3}{16}$ in. were slip-cast and sintered for 2 hr at 1350°C in hydrogen. The physical properties were as follows:

Density	95%
Tensile strength	70,000 psi
Elongation	40.5%

It should be mentioned that, for conventional pressure-compacting of a plate of these dimensions, a 1200-ton press and an expensive hardened steel die would be necessary, whereas for slip casting such a plate, no press, and a plaster of Paris mold costing approximately thirty dollars was used. It is possible to produce inexpensively one or ten plates of this type, but it is possible also to produce in this way hundreds of plates. This illustrates some of the advantages which slip casting has to offer.

What has been shown above for stainless steel is also true for tungsten, molybdenum, cobalt, and other steel powders. It has

Table V. Shrinkage of 316 Stainless Steel Powder,
Slip-Cast and Sintered at 1300°C for 2 hr in Hydrogen

| | *Tensile bar* | | |
	Length in.	*Shoulder diam., in.*	*Reduced section diam. in.*
Dimensions as sintered	4.147	0.505	0.327
% shrinkage	7.89	10.46	13.02
	Flat plate		
	Length, in.	*Width, in.*	*Thickness, in.*
Dimensions as sintered	11.177	1.925	0.147
% shrinkage	8.73	12.30	17.70

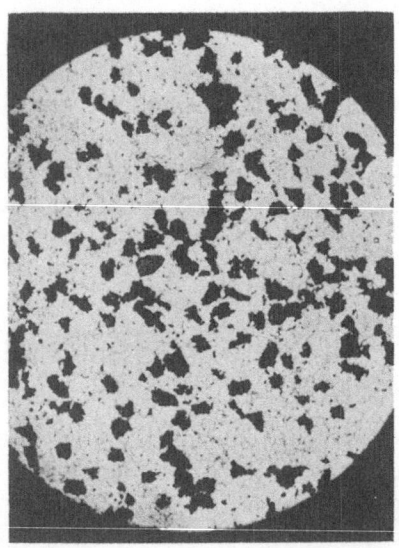

Fig. 6. Structure of slip-cast mixture of 316 stainless steel—UO_2 (75/25 w/o); the oxide (−100 mesh
size) has had −200 mesh size particles eliminated;
sintered at 1350°C for 2 hr in dry hydrogen. Magnification: ×35.

been found, further, that metal−ceramic mixtures of the cermet type can be formed by slip casting, and that it makes very little difference whether the specific gravity of the ceramic component is much greater or lower than that of the metallic component. Experiments were made with slip-cast and sintered stainless steel−aluminum oxide and stainless steel−UO_2 compositions. Slip-cast and sintered stainless steel, mixed with approximately 25 w/o UO_2, showed an elongation of more than 8%, which is quite high for the large amount of the brittle oxide in the stainless steel matrix.

15. A problem of prime importance in powder metallurgy concerns the shrinkage of the pressure-compacted part during sintering. On account of variations in pressure distribution, the shrinkage of pressure-compacted parts is different in the direction of pressing and perpendicular to this direction. Even in each individual direction there are pressure gradients and, therefore, variations in shrinkage. Shrinkage actually is one of the great problems in any powder metallurgy production.

To illustrate the characteristic shrinkage conditions during sintering of compacts made by the slip-casting technique, the dimensional data for a tensile bar specimen and a flat plate are given in Table V. In slip casting also, the shrinkage variations depend upon the dimensions of the part: however, it is most encouraging that the shrinkage along any dimension is perfectly uniform. One may safely state that the shrinkage of slip-cast parts depends on the drying conditions which actually change with the dimensions of the part. However, on account of the absence of pressure and pressure gradients during forming the green compact, the shrinkage of slip-cast parts is, in general, greater but also more uniform in an individual direction than that of pressure-compacted parts.

16. There is a great difference in the structure of sintered materials formed by pressure-compacting and by slip casting, respectively. During pressure-compacting, the powder particles are deformed and strongly cold-worked. New crystal nuclei are formed and recrystallization takes place during sintering.

In the slip-casting process, the powder particles are not deformed and not cold-worked, and practically no nucleation occurs and a continuous grain growth takes place in the mass of powder particles, first within the individual particles and, when diffusion bonding starts, grains may grow from one particle to another. In the slip-cast and sintered material, with a porosity of approximately 5%, one can still recognize the outlines of the original powder particles. This is shown in the photomicrograph, Fig. 6, of a slip-cast and sintered mixture of 75% stainless steel−25% UO_2, at 35 × magnification. This photomicrograph also shows the very uniform distribution of the oxide component in the metal matrix.

17. Slip casting of metal powders and of cermet powder mixtures actually offers a new avenue in powder metallurgy. Slip casting should not, and never will, replace the conventional pressure-compacting of certain parts, but rather permits an extension of metal-powder applications to large and complicated shapes and, therefore, for an economic small-scale production for which the conventional powder metallurgy methods are not applicable.

Slip casting is definitely not just an interesting laboratory method, but offers good aspects for production, and will probably contribute to a considerable growth of the powder metallurgy industry. The data on slip-casting stainless steel presented in this paper are to be taken only as examples. Slip casting of several metals, such as tungsten and molybdenum, is under development, and the results are extremely promising.

Acknowledgment

The work described in this paper has been done under Contract AT(30-1)-1991 with the U. S. Atomic Energy Commission, and a subcontract has been given to Stevens Institute of Technology. The author wishes to thank Professor J. Comstock, Messrs. D. P. Ferriss, E. B. Wilson, and F. W. Heck of the Stevens Institute Powder Metallurgy Laboratory and, further, Messrs. W. G. Lidman and E. N. Mazza of Sylvania-Corning Nuclear Corporation, for their cooperation.

*Chapter 14. Vibratory Compaction**

Vibratory Compacting of Metal Powders [†]

J. L. Brackpool and L. A. Phelps

The B.S.A. Group Research Centre
Birmingham, England

The vibratory compacting of copper powder has been studied using a mechanical vibrator. The major factors influencing the green density of the compacts were the amplitude and frequency of vibration, and the applied pressure. A minimum time of 10 sec on the vibrator was necessary to achieve the maximum density value. Other factors examined were the effects of vibration on blended powders with constituents of widely different densities, and the suitability of this method to compact various materials. Vibratory compacting produced compacts of improved uniformity and green density.

I. Introduction

Vibratory compacting is of special interest in the field of powder metallurgy because it provides the means of compacting metal powders to relatively high densities using low pressures. At high

* Cf. Vibratory Compacting - Principles and Methods (Plenum Press, New York, 1967).
† Contribution to a symposium on Recent Advances in Powder Metallurgy held in London on October 22 and 23 1964.

pressures in direct pressing "layer cracks" are produced as a result of the internal stresses. The low pressures applied in vibratory compacting can also reduce the cost of dies and compacting equipment. This is desirable in small-quantity production and in striving for improved production efficiency.

The method is also applicable to the compacting of nonductile materials [1-4]. Experiments carried out elsewhere have been concentrated on the compaction of uranium oxide, alumina, and various carbides. Likhtman et al. [2], using a mechanical vibrator with a frequency of 230 cps and an amplitude of $1.2 \cdot 10^{-3}$ in., compacted (Ti,W)C and cobalt mixtures to acceptable densities with pressures 100 times less than conventional compacting pressures. Furthermore, uniform mixing and even densities were claimed through the compact. Similarly, Hauth [3], using high-energy electronic vibrators compacted uranium dioxide to densities >90% of theoretical.

The object of the present series of experiments was to find if there was a correlation between powder characteristics and the type of vibrations, by assessing the effects of amplitude, frequency, time, particle size, and other factors on the green density of vibratory-compacted copper.

A second series of experiments covered:

  i. The variation of density throughout a copper block.
  ii. The effect of blending powders of different densities.

It was decided to use a mechanical vibrator because of its low cost, although an electronic vibrator might have given a wider range of control.

II. Apparatus

The vibratory-compacting machine is shown schematically in Fig.1. It consists of a four-strain-rod press with an air cylinder for adjusting the position of the top punch assembly and for applying a known, fixed pressure to the powder bed. The die has a cavity 2 in. deep × 1 in. in diameter and is made from nitrided steel. The vibrator is attached to the bottom punch plate and consists essentially of two contra-rotating shafts with out-of-balance

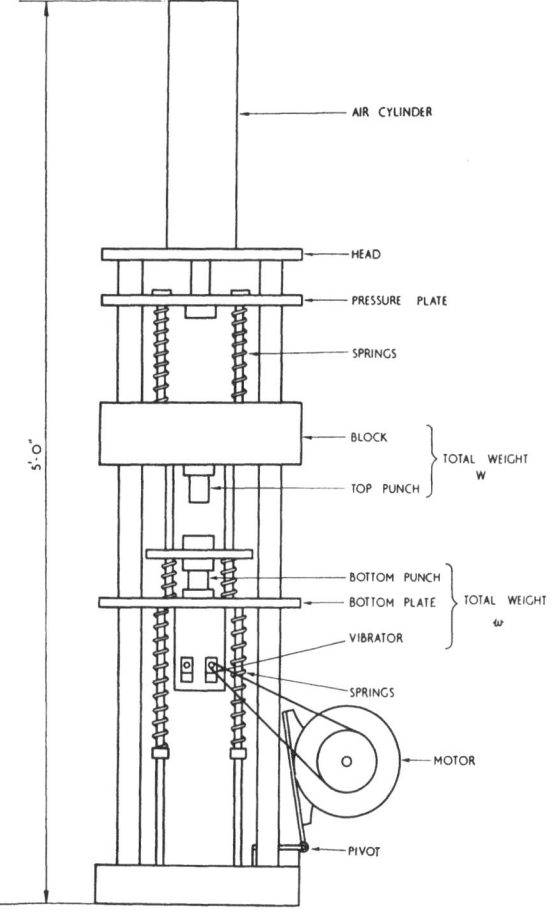

Fig. 1. Schematic layout of vibratory-compacting apparatus.

weights at each end, and is driven by a flexible belt from a $\frac{1}{2}$-hp, dc motor. The out-of-balance weights are positioned on each shaft so that they cancel in the horizontal plane of the vibrator. This arrangement produces vibrations in the vertical direction only.

The vibrations and pressures were applied simultaneously, except where stated in the text, the vibration taking 3 sec to reach a maximum and the pressure 10 sec. The frequency and acceleration were determined from the output signal of an accelerometer,

Table I.   Properties of Electrolytic-
Copper Powder

Particle Size	Wt.-%	Apparent Density, g/cm³	Flow-Rate,* g/sec
−70+100 mesh	6·2		
−100+140 mesh	7·4		
−140+200 mesh	27·4	2·9	1·86
−200+300 mesh	30·3		
−300+400 mesh	18·0		
−400 mesh	9·7		

*Determined according to MPA Standard 3 − 45.

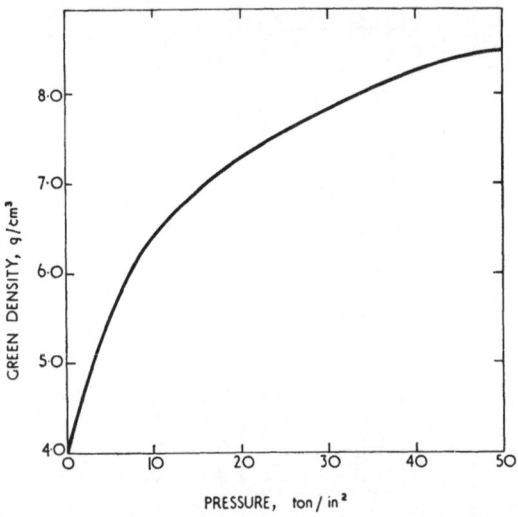

Fig. 2. Pressure/density relationship of as-received copper powder for conventional compacting.

which was amplified by a valve-voltmeter and monitored by a
cathode-ray oscilloscope. The frequency could be varied up to
270 cps with amplitudes of up to ± 0.5 in.

III.   Experimental Details

Preliminary experiments indicated that several factors affected the green density. While some of these factors may apply
only to the present apparatus, each one was given careful consideration. McKechnie's electrolytic-copper powder was used for

Table II. Variation of Density (g/cm^3)
with Time. Amplitude $3 \cdot 3 \cdot 10^{-3}$ in;
pressure, 792 lb/in^2; powder weight, 20 g

Time, sec	Frequency, cps					
	160	170	180	190	200	210
10	$4 \cdot 5_2$	$4 \cdot 8_2$	$5 \cdot 1_0$	$5 \cdot 2_9$	$5 \cdot 4_9$	$5 \cdot 6_2$
15	$4 \cdot 4_8$	$4 \cdot 9_2$	$5 \cdot 1_2$	$5 \cdot 3_3$	$5 \cdot 4_4$	$5 \cdot 7_2$
20	$4 \cdot 5_6$	$4 \cdot 9_3$	$5 \cdot 0_8$	$5 \cdot 3_0$	$5 \cdot 4_5$	$5 \cdot 7_0$
25	$4 \cdot 5_2$	$4 \cdot 8_3$	$5 \cdot 1_1$	$5 \cdot 4_0$	$5 \cdot 5_0$	$5 \cdot 6_4$
30	$4 \cdot 4_9$	$4 \cdot 7_8$	$5 \cdot 1_0$	$5 \cdot 3_8$	$5 \cdot 4_8$	$5 \cdot 6_0$
35	$4 \cdot 5_8$	$4 \cdot 7_0$	$5 \cdot 1_2$	$5 \cdot 3_6$	$5 \cdot 4_5$	$5 \cdot 5_8$
40	$4 \cdot 5_3$	$4 \cdot 9_1$	$5 \cdot 0_0$	$5 \cdot 3_2$	$5 \cdot 4_3$	$5 \cdot 6_5$
80	$4 \cdot 5_7$	$4 \cdot 9_1$	$5 \cdot 1_4$	$5 \cdot 3_2$	$5 \cdot 4_6$	$5 \cdot 6_0$
160	$4 \cdot 5_5$	$4 \cdot 8_2$	$5 \cdot 1_2$	$5 \cdot 3_6$	$5 \cdot 4_8$	$5 \cdot 6_8$

evaluation purposes; the particulars of this powder are given in Table I and Fig. 2. Calculation of the forces involved in vibratory compacting is considered in the Appendix (p. 254).

To obtain a detailed picture of the effect of time on the green density, the time was varied at specified frequencies, amplitudes, and applied pressures, in the range from 10 sec (the time required to reach maximum pressure) to 160 sec under any one set of conditions. A series of density versus time results is given in Table II. These are typical of all the results, from which it is clear that there was no significant variation of green density with time.

The times between the initial application of vibration and the application of pressure had a subsidiary effect on green density, as shown in Table III. The maximum density was reached after a vibration time of 10 sec (before the application of pressure). A further increase in the time of vibration before applying the pressure had no influence on the green density. However, if the pressure reached a maximum before the vibrator was put on, there was a considerable decrease in the green density.

The weight of powder was varied from 5 to 40 g (40 g being the amount of loose powder required to fill the die). The weight to give the optimum green density appeared to be about 20 g (Table IV).

All the compacts produced after the initial experiments were made with the optimum powder weight of 20 g and using a vibra-

Table III. Various Times between Application of Vibration and Pressure and Their Effect on Green Density. Amplitude, $6 \cdot 3 \cdot 10^{-3}$ in.; pressure, 1470 lb/in^2; frequency, 180 cps; powder weight, 20 g

Time between Application of Vibration and Pressure, sec*	Green Density, g/cm^3
25	$6 \cdot 3_7$
20	$6 \cdot 3_6$
15	$6 \cdot 3_7$
10	$6 \cdot 3_8$
7	$6 \cdot 3_6$
6	$6 \cdot 3_5$
5	$6 \cdot 3_2$
4	$6 \cdot 3_3$
3	$6 \cdot 2_1$
0	$5 \cdot 5_3$

* The times have been adjusted to a common datum to allow for the vibrator taking 3 sec and the pressure 10 sec to reach peak values.

Table IV. Effect of Variation of Compact Weight on the Green Density. Amplitude, $3 \cdot 3 \cdot 10^{-3}$ in.; pressure, 1650 lb/in^2; frequency, 180 cps

Weight of Powder, g	Green Density, g/cm^3
5	$5 \cdot 0_9$
10	$5 \cdot 1_0$
20	$5 \cdot 2_9$
40	$4 \cdot 9_6$

tion time of 20 sec, which insured that the maximum density had been achieved.

The influence of amplitude on the green density at various frequencies and pressures was assessed by altering the amount of the out-of-balance weight on the vibrator to give amplitudes of $0.54 \cdot 10^{-3}$, $3.3 \cdot 10^{-3}$, and $6.3 \cdot 10^{-3}$ in. The results of these experiments are shown in Fig. 3.

The pressure to the air cylinder (Fig. 1) was varied from 0 to 100 lb/in.^{2}, resulting in applied pressures on the compact of up to 200 lb/in.^{2}. Figure 4 shows the effect of pressure variation at a constant amplitude on the green density of copper powder over a range of frequencies. Table II, Fig. 3, and Fig. 4 show that the green density increases with frequency. Furthermore, at a given amplitude and pressure there is a particular frequency at which the green density is a maximum.

The influence of particle size on green density is illustrated in Figs. 5 and 6. When the powder had a narrow particle-size distribution the density increased as the mean particle size increased (Fig. 5). It was shown, too, that at a constant mean particle size, the green density was increased both by pressure and frequency. Adding increasing percentages of "fine" powder (mean particle size 54 μ) to

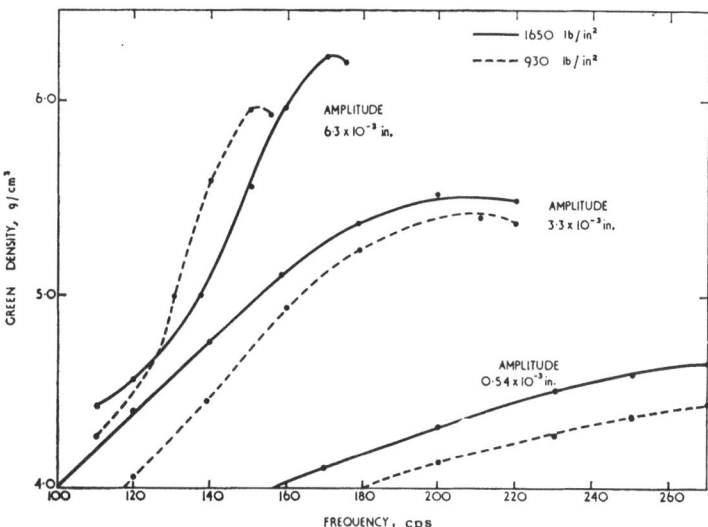

Fig. 3. Effect of variation in amplitude on the density of copper.

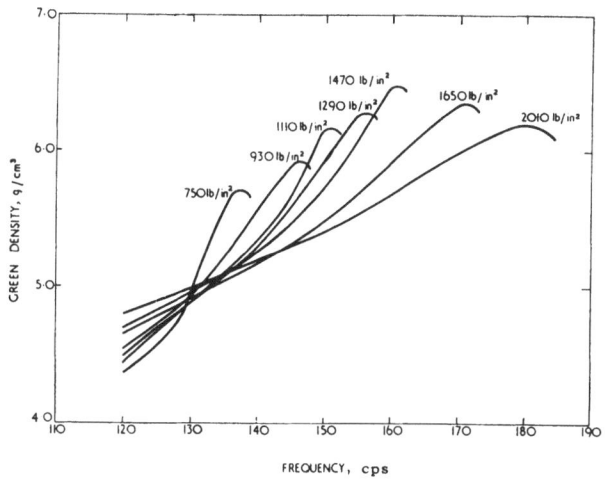

Fig. 4. Effect of variation in applied pressure on the density of copper at varying frequencies. Amplitude 6.3 · 10^{-3} in.

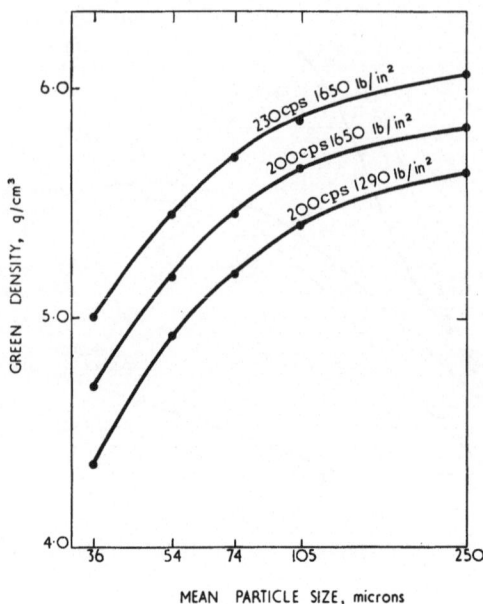

Fig. 5. Effect of particle size on the density of electro-lytic-copper powder compacts. Amplitude $3.3 \cdot 10^{-3}$ in.

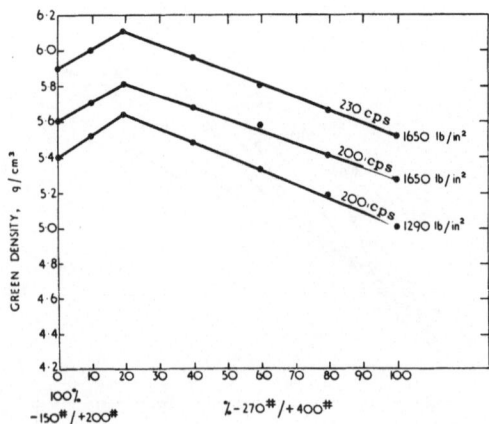

Fig. 6. Effect of additions of "fine" powder to "coarse" powder on the density of electrolytic-copper powder compacts. Amplitude $3.3 \cdot 10^{-3}$ in.

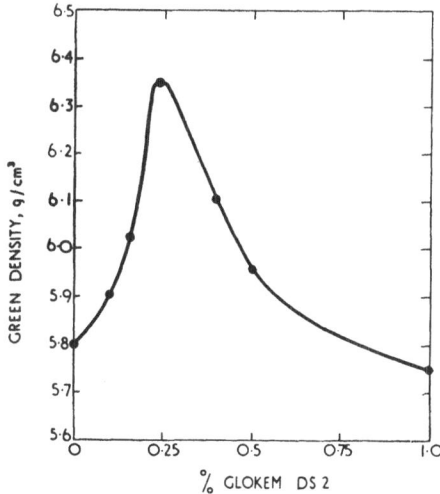

Fig. 7. Effect of an additive on the density of copper powder.   Amplitude 3.3 · 10^{-3} in; frequency
150 cps; pressure 1290 lb/in.^{2}.

"coarse" powder (mean particle size 105 μ) resulted in an increase
in green density to a maximum at a ratio fine : coarse of 1 : 4 (Fig.
6).

Various additives were blended with copper powder to find
their effect on the green density.   These included lithium stearate,
Abril SS Wax, and Glokem DS2.   Figure 7 shows the effect of adding Glokem DS2 to "as-received" copper powder compacted at
1290 lb/in.^{2} with a frequency of 150 cps and an amplitude of 3.3
10^{-3} in.

Two methods were used to determine the variation of green
density throughout the copper compacts.   The first method involved sectioning a compact into small pieces ($\sim\frac{1}{8}$-in. cubes) and
measuring their density by the Archimedes method, after coating
the pieces with lacquer to prevent infiltration of water.   Alterna-
tively, a hardness survey was made of the compact, and the correlation of hardness and green density was obtained by separate
experiments.   The density variation in compacts produced under
varying conditions and using varying particle size was determined
by these methods and is illustrated in Fig. 8.

Table V. Densities of Vibratory-Compacted Powders

Powder	%Theoretical Density	Particle Size	Amplitude, in.	Pressure, lb/in²	Frequency, cps
Copper	75·5	Maximum-density blend	6·3 × 10⁻³	1470	160
Silicon	75·3	−200 mesh	3·3 × 10⁻³	1650	140
Alumina	68·3	−300 mesh	3·3 × 10⁻³	1650	135
Stainless Steel	48·0	−60 mesh	3·3 × 10⁻³	1470	165
Tungsten	N.C.*	5 μ	—	—	—

*Did not compact under any conditions.

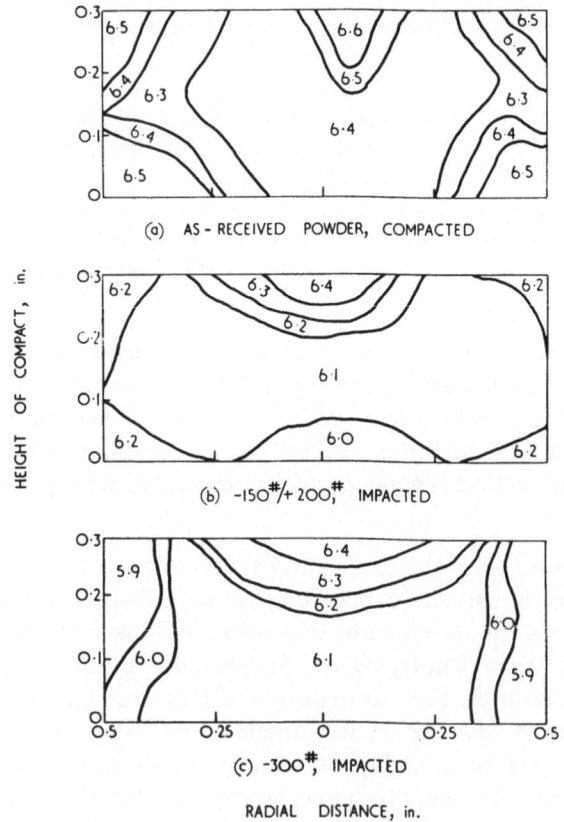

HEIGHT OF COMPACT, in.

(a) AS - RECEIVED POWDER, COMPACTED

(b) -150#/+ 200;# IMPACTED

(c) -300#, IMPACTED

RADIAL   DISTANCE, in.

Fig. 8. Variation in green density of vibratory-compacted or vibratory-impacted copper powder (see Section IV).

Experiments were carried out to assess the effect of vibratory-compacting on the segregation of particles of widely differing densities. Several mixtures of copper and tungsten powders of similar particle-size ranges were compacted under various conditions and were then examined metallographically to show the distribution of the tungsten particles. There was little or no segregation of the two powders. The majority of the tungsten particles were completely surrounded by copper, voids being present only in the copper matrix.

A variety of materials was examined to study the suitability of the apparatus to compact powders with different pressing characteristics. Some of the results obtained are given in Table V.

IV. Discussion of Results

Copper was used to evaluate the apparatus because it was relatively easy to compact it to suitable green densities under varied conditions. It was found from these experiments that two distinct mechanisms operated in the vibratory-compaction process. The first took place when the applied force was greater than the vibratory force,

$$F_p > F_v \tag{1}$$

and the second when the applied force was equal to, or less than, the vibratory force

$$F_p \leqslant F_v \tag{2}$$

The latter condition was termed "impacting," since the punch left the compact every cycle and rebounded with an impacting force. The onset of "impacting" was characterized by a sharp increase in the green density of the compacts.

From the results involving time variation it can be seen that ~10 sec was necessary to achieve maximum packing of the particles and after this period, no further densification took place. However, there appeared to be no simple explanation for this phenomenon. Undoubtedly, a certain minimum time would be expected to be required for packing, but it was thought that frequency and amplitude would have had some effect on this time. This was not borne out by our findings.

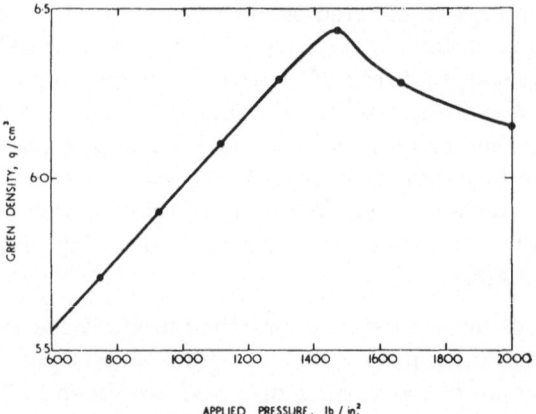

Fig. 9. Effect of variation in the applied pressure on the peak density of vibratory-compacted copper powder. Amplitude, $6.3 \cdot 10^{-3}$ in.

Frequency and amplitude both increased the green density by increasing the vibratory force applied to the powder compact. Frequency had a much greater influence at higher amplitudes. These higher amplitudes may have been necessary to overcome the frictional forces between particles.

An increase in applied pressure at low frequencies increased the green density; on the other hand, as the "impacting" range was approached, low pressures produced higher green densities; for example, the maximum green density obtained with an amplitude of $3.3 \cdot 10^{-3}$ in. on the "as-received" copper powder was at an applied pressure of ~150 lb/in.^{2} at 160 cps. Packing and densification occurred simultaneously only over a limited time period. After a certain pressure had been reached, rearrangement of the free particles ceased and further densification occurred by deformation of the copper powder. Thus, as the applied pressure was raised to 1470 lb/in.^{2}, the maximum green density obtained increased; further increases in pressure caused a decrease in the final density (Fig. 9). This was because the most favorable arrangement of particles was not reached, with a shorter period of vibration, before the particles were consolidated.

In the densification of nonductile materials, deformation played only a minor role. That is why the green density was a small fraction of the theoretical density in vibratory-compacted alumina and carbides as compared with copper powder. This is substantiated to some extent by the work of Likhtman et al. [2] on mixed carbides, where only relatively small pressures were required to give maximum densities [7].

The addition of an increasing percentage of fines to the coarse powder gave a maximum green density at a ratio of 1 : 4; this is in good agreement with the work of Carman [5]. In a recent paper, Evans and Millman [6] reported that maximum packing occurred with three modal sizes; experiments on compacting with this type of distribution are in progress.

The variations in green density observed in compacts produced by vibratory compacting were much less than those found in compacts produced in double-acting static dies. This is of major importance in component production, where shrinkage on sintering not only has to be a minimum, but should be uniform. The "as-received" powder, having a wider range of particle sizes, produced higher green densities than sieved fractions of powder, as was to be expected. The lowest green densities were observed with the −300-mesh powder. The difference in density distribution between fine and coarse powder fractions could be associated with the surface area and the inertia of the powder particles. The former would affect the heating and, therefore, the recovery from cold work in the individual particles, while the latter would affect the ability of the powder particles to form a fluidized bed in the die.

The experiments using mixtures of copper and tungsten powder showed that there was little or no segregation of the two powders. Furthermore, the majority of the tungsten particles were completely surrounded by copper, any voids present being only in the copper matrix. This could be used to advantage in the production of compacts of intermetallics and nonmetallics in a ductile matrix, e.g., in the fabrication of both dispersion-strengthened alloys and superconducting alloys.

V.   Conclusions

Two distinct conditions govern the vibratory-compacting process:

a. When the applied force is greater than the vibratory force, namely the "compacting" condition.

b. When the applied force is equal to, or less than, the vibratory force, namely, the "impacting" condition.

The onset of "impacting" is associated with a marked increase in the green density of compacts.

A study of the effects of the different processing variables enables the following conclusions to be drawn:

1. A minimum time of 10 sec is necessary to achieve the maximum packing and densification of a loose powder with vibration.

2. Frequency and amplitude both increase the green density as a result of increasing the applied vibratory force.

3. At low frequencies the green density increases with the pressure, but at higher frequencies the green density increases as the pressure decreases.

4. The particle-size distribution has a strong influence on the green density.

5. Mixtures of powders of widely differing densities showed little or no segregation after vibratory compacting.

6. Vibratory compacting produced compacts of improved uniformity and green density.

Acknowledgments

The authors would like to thank their colleague, Mr. C. F. May, for his help in the calculation of the vibratory and impacting forces, given in the Appendix. The whole of this work has been coordinated by Mr. D.A. Oliver, C. B. E. , M. Sc. , F. Inst. P. , Director of Research. The B. S.A. Group Research Centre, whose encouragement is greatly valued.

Appendix

1. Basic Principles

The arrangement of the vibratory-compacting press is shown in Fig. 1. The pneumatic cylinder (in combination with the weight of the top assembly W) provides a direct compacting force F_p, while the rotary vibrator exerts a vertical cyclic force $f_V = F_V \sin \omega t$ on the masses to which it is coupled.

The two principal masses in each case have been designated W and w (Fig. 1) and it is desirable that W should be large compared with w.

Coupling between these two masses takes place through: (a) the powder compact and (b) the top and bottom springs and the air in the cylinder. The coupling (b) will be of low stiffness because of the compressibility of the pressurized air in the cylinder and the low stiffness of the springs. Hence, the transmission of vibratory forces between the two masses via this coupling will be small and can be neglected.

The coupling (a), through the powder, will be spongy initially, but as compaction takes place the coupling will become stiff and capable of transmitting large vibratory forces. If in this latter condition the two masses W and w are regarded as one combined mass W + w which is subjected to a force from the vibrator of $f_V = F_V \sin \omega t$, then the acceleration of the combined mass is

$$\frac{d^2 \chi}{dt^2} = \frac{g}{W+w} F_v \sin \omega t \tag{3}$$

and the displacement from the mean position

$$\chi = \frac{g}{W+w} \iint F_v \sin \omega t$$

$$= \frac{g}{W+w} \frac{F_v}{\omega^2} \sin \omega t \tag{4}$$

from which the total or peak-to-peak vibratory displacement is

$$D = \frac{2g}{W+w} \frac{F_v}{\omega^2} \tag{5}$$

The force required to accelerate the mass W, according to (3) above is

$$f_s = \frac{W}{g} \frac{g}{W+w} F_v \sin \omega t$$

$$= \frac{W}{W+w} F_v \sin \omega t \tag{6}$$

and since this force is transmitted to the mass W via the compact, f_C will be the vibratory-compacting force which will have a peak value of

$$F_c = \pm \frac{W}{W+w} F_v$$

(7)

The advantage of making W large compared with w will be apparent from Eq. (7).

2. Impacting

The foregoing will apply, of course, only if the steady direct-compacting force F_p is greater than the peak vibratory-compacting force F_C, since the compact can transmit only compressive forces. If F_C becomes $> F_p$, then the punches will separate from the compact and return to it with an impact at one part of each cycle. The transient forces set up by such impacts are not readily calculable, but they will obviously have a much greater magnitude than the corresponding controlled cyclic forces existing if "impacting" is prevented by increasing F_p to a value in excess of F_C.

3. Force Produced by the Vibrator

The vibrator has two contra-rotating shafts, and each shaft has an unbalanced mass w_V at each end. If the radius of rotation of the center of gravity of the unbalanced masses is r, then the vertical acceleration of each mass is $\omega^2 r \sin \omega t$, and the force produced by the vibrator will be

$$f_v = \frac{4w_v}{g} \omega^2 r \sin \omega t$$
$$= F_v \sin \omega t$$

(8)

hence,

$$F_v = \frac{4w_v}{g} \omega^2 r$$

(9)

Substituting Eq. (9) in Eq. (5) gives the vibratory displacement

$$D = \frac{2g}{W+w} \frac{1}{\omega^2} \frac{4w_v}{g} \omega^2 r$$
$$= \frac{8w_v r}{W+w}$$

(10)

Therefore, D is independent of the speed of rotation of the vibrator ω.

Substituting Eq. (9) in Eq. (7) gives the peak vibratory-compacting force

$$F_c = \pm \frac{W}{W+w} \frac{4w_v}{g} \omega^2 r$$

(11)

Equations (10) and (11) show that if W is made increasingly large, $D \rightarrow 0$ and F_c approaches a finite value equal to $(4w_v/g)\omega^2 r$. In practice, W will be relatively large and hence D will be small in the final stages of compacting. This suggests that the bulk vibratory movement is irrelevant in the final stages of the process and that any benefits obtained are due to the cyclic variations in the compressive force applied by the dies. For this reason F_c (rather than D) would seem to be the significant parameter against which to compare results.

In the initial stages of the process there may well be a "cocktail-shaking" effect, particularly if the vibration is applied before the dies close in on the powder, but this action probably ceases at an early stage when the individual particles become locked together. The exact nature of the subsequent compacting action is one which requires detailed investigation.

References

1. W. C. Bell, R. C. Dillender, H. R. Lominac, and E. G. Manning, J. Am. Ceram. Soc., 38 : 396 (1955).
2. V. I. Likhtman, N. S. Gorbunov, I. G. Shatalova, and P. A. Rebinder, Dokl. Akad. Nauk SSSR, 134 : 1150 (1960).
3. V. V. Hauth, U. S. Atomic Energy Commission Rept. (HW 67777) (1961).
4. I. Sheinhartz and J. Fugardi, Sylvania-Corning Nuclear Corpn. Rept. (SCNC-311) (1960).
5. P. C. Carman, Trans. Inst. Chem. Eng., 15 : 150 (1937); 16 : 168 (1938).
6. P. E. Evans and R. S. Millman, Powder Met., No. 13 : 50 (1964).
7. N. S. Gorbunov, I. G. Shatalova, V. I. Likhtman, N. V. Mikhailov, and P. A. Rebinder, Poroshkovaya Met., 1 (6) : 10 (1961).